Tropical Medicine

EDITED BY

DION R. BELL

MB ChB (Leeds) FFPHM FRCP DTM & H (Liverpool)
Formerly Reader in Tropical Medicine
Liverpool School of Tropical Medicine

Fourth edition

b

**Blackwell
Science**

© 1981, 1985, 1990, 1995 by
Blackwell Science Ltd
Editorial Offices:
Osney Mead, Oxford OX2 0EL
25 John Street, London WC1N 2BL
23 Ainslie Place, Edinburgh EH3 6AJ
350 Main Street, Malden
  MA 02148 5018, USA
54 University Street, Carlton
  Victoria 3053, Australia
10, rue Casimir Delavigne
  75006 Paris, France

Other Editorial Offices:
Blackwell Wissenschafts-Verlag GmbH
Kurfürstendamm 57
10707 Berlin, Germany

Blackwell Science KK
MG Kodenmacho Building
7–10 Kodenmacho Nihombashi
Chuo-ku, Tokyo 104, Japan

First published 1981
Second edition 1985
Reprinted with corrections 1987
Third edition 1990
Reprinted 1991, 1992, 1993
Fourth edition 1995
Reprinted 1996 (twice), 1997, 1998

Set by Excel Typesetters, Hong Kong
Printed and bound in Great Britain
at the University Press, Cambridge.

The Blackwell Science logo is a
trade mark of Blackwell Science Ltd,
registered at the United Kingdom
Trade Marks Registry

For further information on
Blackwell Science, visit our website:
www.blackwell-science.com

DISTRIBUTORS

Marston Book Services Ltd
PO Box 269
Abingdon, Oxon OX14 4YN
(Orders: Tel: 01235 465500
        Fax: 01235 465555)

USA
Blackwell Science, Inc.
Commerce Place
350 Main Street
Malden, MA 02148 5018
(Orders: Tel: 800 759 6102
         781 388 8250
    Fax: 781 388 8255)

Canada
Login Brothers Book Company
324 Saulteaux Crescent
Winnipeg, Manitoba R3J 3T2
(Orders: Tel: 204 224-4068)

Australia
Blackwell Science Pty Ltd
54 University Street
Carlton, Victoria 3053
(Orders: Tel: 3 9347 0300
    Fax: 3 9347 5001)

A catalogue record for this title
is available from the British Library

ISBN 0-632-03839-X

Library of Congress
Cataloging-in-Publication Data

Lecture notes on tropical medicine/
  edited by Dion R. Bell. – 4th ed.
        p.   cm.
    2nd ed. cataloged under: Bell, Dion R.
    Includes bibliographical references
    and index.
    ISBN 0-632-03839-X
    I. Tropical medicine – Handbooks, manuals,
    etc.
    I. Bell, Dion R.
  RC961.B396 – 1995
  616.9'883 – dc20   94-33311
                                    CIP

Contents

Contributors, vii
Preface to the fourth edition, ix
Preface to the first edition, ix
List of abbreviations, xi

Section 1: Diseases commonly presenting as fevers
1 Malaria, 3
2 Visceral leishmaniasis, 38
3 African trypanosomiasis, 50
4 South American trypanosomiasis (Chagas' disease), 67
5 Relapsing fever, 70
6 Rickettsial infections, 73
7 Arbovirus and rodent-borne infections, and rabies, 77
8 Typhoid and paratyphoid fevers (enteric fevers), 94
9 Tuberculosis in the tropics, 102
10 HIV infection and disease in the tropics, 113
11 Fevers in general, 130

Section 2: Diseases commonly presenting as diarrhoea
12 Amoebiasis, 141
13 Giardiasis, 156
14 Balantidiasis, 161
15 Cholera, 162
16 Diarrhoea in general, 169

Section 3: Soil-transmitted helminths
17 *Ascaris lumbricoides*: the large round worm, 181
18 Hookworm, 185
19 *Trichuris (Trichocephalus) trichiura*: the whip worm, 194
20 *Strongyloides stercoralis*, 197
21 *Toxocara canis* (and *cati*), 202

Section 4: Tapeworms
22 *Taenia saginata*: the beef tapeworm, 209
23 *Taenia solium*: the pork tapeworm, 211
24 Hydatid disease: *Echinococcus* infections, 215

Section 5: Flukes
25 Blood flukes: schistosomes, 221
26 Oriental liver flukes, 240
27 Lung flukes, 243
28 Intestinal flukes, 246

Section 6: The filarial worms
29 *Wuchereria bancrofti*, 251
30 *Brugia malayi*, 258
31 *Loa loa*, 260
32 *Onchocerca volvulus*, 263
33 *Dracunculus medinensis*: guinea worm infection, 271
34 Tropical eosinophilia syndrome, 275

Section 7: Anaemia in the tropics
35 Anaemia in the tropics, 279

Section 8: Leprosy
36 Leprosy, 293

Section 9: Tropical skin conditions
37 Tropical ulcers, 313
38 Buruli ulcer, 317
39 Cutaneous leishmaniasis, 319
40 Ulcers in general, 327
41 Itchy skin lesions in the tropics, 328
42 Miscellaneous exotic conditions, 330

Section 10: Snake Bite
43 Snake bite, 335

Appendix—Answers, 341

Index, 355

Contributors

RAY FOX
MB MRCP DTM & H
Clinical Lecturer, School of Tropical Medicine
Pembroke Place, Liverpool L3 5QA
[chapter 9]

CHARLES GILKS
MB MRCP DPhil DTM & H
Senior Lecturer, School of Tropical Medicine
Pembroke Place, Liverpool L3 5QA
[chapter 10]

MALCOLM MOLYNEUX
MD FRCP
Professor, School of Tropical Medicine
Pembroke Place, Liverpool L3 5QA
[chapters 1, 11–14, 25–28, 35]

DAVID SMITH
MB FRCP DTM & H
Senior Lecturer, School of Tropical Medicine
Pembroke Place, Liverpool L3 5QA
[chapters 2–4, 7, 15, 37–42]

DAVID THEAKSTON
BSc PhD
Senior Lecturer, School of Tropical Medicine
Pembroke Place, Liverpool L3 5QA
[chapter 43]

GEORGE B. WYATT
FRCP FFCM FFPHM
Senior Lecturer, School of Tropical Medicine
Pembroke Place, Liverpool L3 5QA
[chapters 5, 6, 16–24, 29–34, 36]

Preface to the fourth edition

The fourth edition is different from the previous three because I have enlisted the help of colleagues from the Liverpool School of Tropical Medicine to bring it up to date. So the book should be less idiosyncratic than its predecessors, and benefit from cumulative experience far greater than my own. I have acted as editor, so I hope continuity will be evident.

The purpose of the book remains unaltered: to give practical help to doctors and other health workers encountering problems with tropical diseases, especially in developing countries. We also aim to give sound guidance on the solution of these problems in more sophisticated settings.

As with the third edition, all the royalties from the sale of this book are used to support medical students spending their electives in tropical countries.

Dion R. Bell
York 1994

Preface to the first edition

This is not meant to be a textbook of tropical medicine, because it is not comprehensive and contains very few references. It is also unbalanced, for the attention given to different topics does not reflect their prominence in the world literature.

My aim has been to present most comprehensively those conditions which cause serious disease in a significant number of people, and which are amenable to effective treatment or control. The main objective is to give help to those who have to face practical problems in tropical medicine, wherever they may be.

Dion R. Bell
Liverpool 1981

List of abbreviations

AFB	acid-fast bacilli
AIDS	acquired immunodeficiency syndrome
ALA	amoebic liver abscess
ARC	AIDS-related complex
BB	borderline borderline leprosy
BCG	bacillus Calmette-Guérin
BL	borderline leprosy
BT	borderline tuberculoid leprosy
CATT	card agglutination test for trypanosomiasis
CFT	complement fixation test
CHR	*cercarien Hüllen Reaktion*
CIE	countercurrent immunoelectrophoresis
CL	cutaneous leishmaniasis
CMI	cell-mediated immunity
CNS	central nervous system
COP	circumoval precipitin
CSF	cerebrospinal fluid
CT	computerised tomography
DAT	direct agglutination test
DCL	disseminated cutaneous leishmaniasis
DDT	dichlorodiphenyltrichloroethane
DEC	diethylcarbamazine
DFMO	difluoromethyl ornithine
DHF	dengue haemorrhagic fever
DIC	disseminated intravascular coagulation
DSS	dengue shock syndrome
ECG	electrocardiogram
EE	exoerythrocytic
ELISA	enzyme-linked immunosorbent assay
ENL	erythema nodosum leprosum
ESR	erythrocyte sedimentation rate
ETEC	enterotoxigenic strains of *Escherichia coli*
G6PD	glucose-6-phosphate dehydrogenase deficiency
GPA	Global Programme on AIDS
Hb	haemoglobin
HbSC	sickle-cell haemoglobin C
HbSS	sickle-cell haemoglobin
HDCV	human diploid cell vaccine
HIV	human immunodeficiency virus

HLA	human leucocyte antigen
HWA	hookwork anaemia
IFAT	indirect fluorescent antibody test
IgG	immunoglobulin G
JVP	jugular venous pressure
LE	lupus erythematosus
LL	lepromatous leprosy
MAEC	minianion exchange column
MBL	multibacillary leprosy
MCH	mean corpuscular haemoglobin
MCHC	mean corpuscular haemoglobin concentration
MCV	mean corpuscular volume
mf	microfilariae
MHCT	microhaematocrit
MIC	minimum inhibitory concentration
NTS	non-typhi salmonellae
PBL	paucibacillary leprosy
PCECV	purified chick embryo cell vaccine
PCR	polymerase chain reaction
PCV	packed cell volume
PE	pre-erythrocytic
PGL	persistent generalized lymphadenopathy
PNL	polymorphonuclear leucocytosis
PPD	purified protein derivative (tuberculin antigen)
PUO	pyrexia of undetermined origin
PVCV	purified vero cell vaccine
QBC	quantitative buffy coat
RBC	red blood cell count
RE	reticuloendothelial
RI	reticulocyte index
SBE	subacute bacterial endocarditis
SIV	simian immunodeficiency virus
SLE	systemic lupus erythematosus
SPAG	sensitized particle agglutination test
STD	sexually transmitted disease
TB	tuberculosis
TCBS	thiosulphate–citrate–bile salt–sucrose
TCE	tetrachloroethylene
TIBC	total iron-binding capacity
TIF	thiomersal, iodine, formol
TrU	tropical ulcer
TU	tuberculin units
VLM	visceral larva migrans
VSO	Voluntary Services Overseas
WBC	white blood cell count
WHO	World Health Organization

Diseases commonly presenting as fevers

Malaria

IMPORTANCE AND DISTRIBUTION

Malaria is the most important of all tropical diseases, causing many deaths and much morbidity. It is widely distributed in the tropical and subtropical zones (see Fig. 1.1: this is not necessarily accurate as the world is changing very rapidly). There are four parasite species that cause human malaria, all of which belong to the genus *Plasmodium*:

1 *P. falciparum* (malignant tertian malaria, subtertian malaria).
2 *P. vivax* (benign tertian malaria).
3 *P. ovale* (ovale tertian malaria).
4 *P. malariae* (quartan malaria).

LIFE CYCLE (FIG. 1.2)

Immediately after infection

Malaria is usually transmitted by the bite of an infected female anopheline mosquito. The infecting agent is the sporozoite, a microscopic spindle-shaped cell which is in the mosquito's saliva. Thousands of sporozoites may be injected in a single bite. Infection may also be acquired transplacentally and by blood transfusion or inoculation, via the blood stages of the parasite. The sporozoites disappear from the blood within 8 h, and the successful ones enter polygonal liver cells (hepatocytes). Inside the liver cell the sporozoite divides by asexual fission to form a cyst-like structure called a pre-erythrocytic (PE) schizont, which contains thousands of merozoites. Each merozoite consists of a small mass of nuclear chromatin within a tiny sphere of cytoplasm. The process by which the malaria parasites multiply asexually is called schizogony, whether it takes place in a hepatocyte or in an erythrocyte.

Tissue schizogony

When the PE schizont is mature, it ruptures and liberates its contained

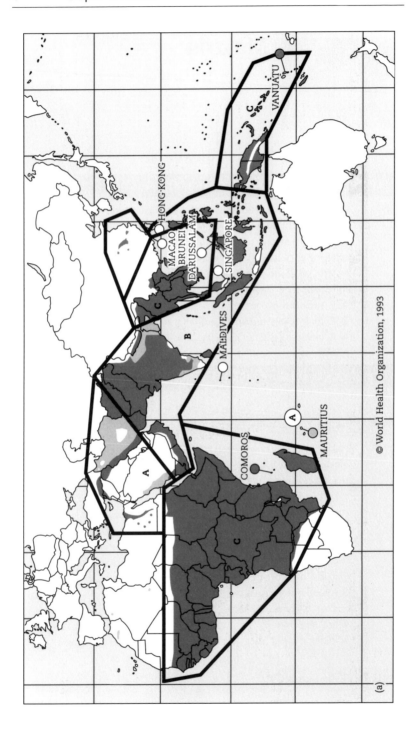

© World Health Organization, 1993

(a)

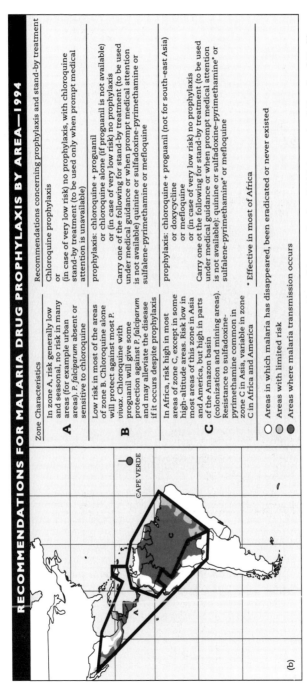

Fig. 1.1 Recommendations for malaria drug prophylaxis by area (1994). From World Health Organization (1994) *International Travel and Health*, WHO, Geneva.

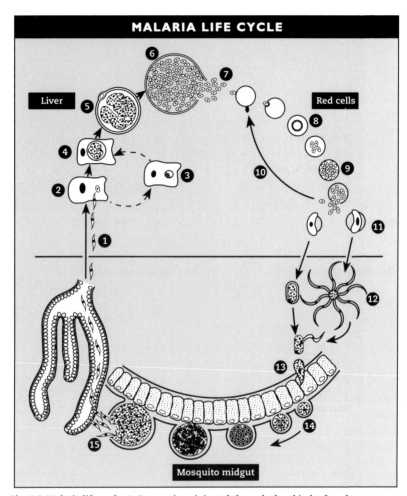

Fig. 1.2 Malaria life cycle. 1, Sporozoites, injected through the skin by female anopheline mosquito; 2, sporozoites infect hepatocytes; 3, some sporozoites develop into 'hypnozoites' (P. vivax and P. ovale only); 4, liver-stage parasite develops; 5–6, tissue schizogony; 7, merozoites are released into the circulation; 8, ring-stage trophozoites in red cells; 9, erythrocytic schizogony; 10, merozoites invade other red cells; 11, some parasites develop into female (macro-) or male (micro-) gametocytes, taken up by mosquito; 12, mature macrogametocyte and exflagellating microgametes; 13, ookinete penetrates gut wall; 14, development of oocyst; 15, sporozoites penetrate salivary glands. Reproduced with permission from Zaman, V., *Atlas of Medical Parasitology*. (1978) Published by ADIS Press, Balgawlah, NSW, Australia.

merozoites, these now enter the blood stream and, in the presence of suitable surface receptors, attach themselves to red cells. When they enter the red cells, the process of blood schizogony begins.

It used to be believed that the tendency of patients with *P. vivax* and *P. ovale* malaria to relapse after the blood stages of the parasites had been eliminated was caused by a persistent cycle in the liver. It was thought that some merozoites from the PE schizonts reinvaded liver cells, and that infection could take place from liver cell to liver cell in this way. These second and subsequent generations of liver schizonts were called exoerythrocytic (EE) schizonts. But no one has ever been able to demonstrate them, and the phenomenon of relapse is now explained by the concept of the hypnozoite. So tissue schizogony nowadays is believed to be confined to the initial cycle of multiplication in the liver, in the form of the PE schizont only.

Prepatent period and relapse: the hypnozoite concept

The time between the bite of the infecting mosquito and the appearance of parasites in the blood is the prepatent period. It is 7–30 days in *P. falciparum* (usually around 10 days), and longer in the other species. It may be very long—in the case of *P. vivax* and *P. ovale* many months or even more than a year. This is believed to be due to the dormant stage of the parasite in the liver. It is as if the sporozoite enters a liver cell and promptly goes to sleep. This dormant stage of the parasite is the hypnozoite. But the dormant parasite has a biological alarm clock, which wakes it from dormancy at a predetermined time. Some strains of parasite 'sleep' longer than others.

The hypnozoite concept explains both a prolonged incubation period and the phenomenon of relapse. The biological 'alarm clocks' of the hypnozoites are set at different times. When each wakes from dormancy the blood is reinvaded, and the patient suffers from a relapse.

Blood schizogony

Once in the circulation, all species of *Plasmodium* multiply by asexual multiplication in erythrocytes—blood schizogony (versus tissue schizogony). After entering the red cell, the merozoite begins to feed on the red cell contents, and because it begins to grow, is now called a trophozoite. Feeding is by ingestion of red-cell stroma and its digestion in a food vacuole. Digested haemoglobin gives rise to a characteristic pigment, malaria pigment (haemozoin), which is present in the cell in increasing amounts as the trophozoite becomes mature. The mature trophozoite begins to divide into separate merozoites within 1–3 days depending on the species, and this process of schizogony is completed in 48h in the case of *P. falciparum*, *P. vivax* and *P. ovale* and 72h in the case of *P. malariae*. When fully developed, the schizont ruptures the red cell containing it, and liberates the merozoites into the circulation. These merozoites will then enter new red cells, and this process of asexual replication in the blood tends to proceed, at least in the early stages, in a logarithmic manner. The parasitaemia (the proportion of red cells containing parasites) never increases as rapidly as the number of merozoites in the red cells would suggest, indicating that not all the merozoites succeed in infecting new red cells, or alternatively that not all trophozoites manage to proceed to schizogony and to the release of merozoites.

Site and periodicity of schizogony

Schizogony occurs in the circulating blood in the cases of *P. vivax*, *P. ovale* and *P. malariae*, so in all these infections schizonts are commonly seen in the peripheral blood films from infected patients.

In *P. falciparum* schizogony only occurs in capillaries deep within the body. At the stage of the maturing trophozoite, parasite antigens are expressed on the surface of the red cell. Some of these antigens are capable of linking to receptors expressed on the endothelial cells lining capillaries in various organs and tissues of the body. The resulting cytoadherence of parasitized erythrocytes to endothelial surfaces leads to the gathering, or sequestration, of large numbers of mature parasites in deep tissues. As a result, schizonts of *P. falciparum* are seldom found in peripheral blood films, and when they are it is usually in patients with very severe infections or after splenectomy. Sequestration probably affords some advantage to the parasite, in that sequestered parasites are no longer available for destruction by the spleen. Sequestration of enormous numbers of parasites may be responsible for some of the severe manifestations of the disease, such as cerebral malaria, which occur only in *P. falciparum* infections.

The periodicity of schizogony characteristically coincides with paroxysms of fever and this led to the traditional names of the different types of human malaria:

1 Tertian malaria (fever every third day, if the first day is given the number 1): *P. vivax* and *P. ovale*.
2 Subtertian malaria (fever slightly more often than every third day): *P. falciparum*.
3 Quartan malaria (fever every fourth day if the first day is given the number 1): *P. malariae*.

   P. falciparum malaria was sometimes called malignant tertian malaria, because of its much greater lethal potential than the other tertian malarias. These antique names for malaria are best avoided, not only because they can be confusing, but because the periodicity they imply often fails to develop. There are many patients who have lost their lives from *P. falciparum* malaria because they never developed the periodic fever which their doctors wrongly believed to be invariable.

SEXUAL CYCLE

Some of the merozoites entering red cells do not develop into schizonts, but develop more slowly into solid-looking parasites called gametocytes. These may persist in the circulation for many weeks without destroying the red cells containing them, and they are the forms infective to the mosquito. In each species of malaria, the gametocytes are differentiated into male and female. When the female mosquito swallows the male and female gametocytes in her blood meal, they develop further in her stomach. The male gametocytes rapidly develop to produce sperma-tozoon-like microgametes, and the female gametocyte becomes the egg-like macrogamete.

Fertilization takes place when a microgamete unites with the macrogamete, and a motile zygote, the ookinete, is produced by this union. The ookinete penetrates to the outer surface of the mosquito's stomach and there develops into an oocyst, which comes to contain thousands of sporozoites. When mature, the oocyst ruptures and liberates the sporozoites into the mosquito's body cavity. The sporozoites then migrate forwards to the salivary glands, and are then ready to infect another victim when the mosquito bites. The time elapsing between the ingestion of the gametocytes and the saliva of the mosquito becoming infective by containing sporozoites is called the extrinsic incubation period. It is variable in the different species of parasite, with different mosquito vectors, and with environmental factors, especially temperature. It is never shorter than 10 days and often much longer.

CLINICAL FEATURES

Infections with all the four different malaria species have many clinical features in common. These are related to the liberation of fever-producing substances, especially during schizogony, and the fact that every red cell containing a trophozoite will be destroyed within 48–72 h. The common features are:

I *Fever*: often irregular. Fever is believed to be mediated by host cytokines, which are secreted by leucocytes and other cells in response to a pyrogen or toxin released by rupturing schizonts.

The pattern of regularly periodic fever often does not occur until the illness has continued for a week or more. It depends on synchronized schizogony. Why schizogony should ever become synchronized is unknown, but an intriguing explanation has been suggested. High temperatures slow the growth of mature, more than of young, parasites. Fever itself may therefore allow young parasites to 'catch up' with older ones, leading to increasing synchrony with successive cycles.

2 *Anaemia.* This is haemolytic in type. It is usually most severe in *P. falciparum* because in this infection cells of all ages can be invaded, and even unparasitized red cells may undergo haemolysis. Also, the parasitaemia in this infection can be much higher than in other malarias.

One would expect a brisk reticulocytosis in a haemolytic anaemia, but this is often absent in acute malaria, presumably because of marrow suppression, which is another recognized effect of circulating cytokines. Marrow aspirates often show evidence of dyserythropoiesis, including phagocytosis of parasitized red cells by macrophages, and phagocytosis of apparently uninfected red cells.

3 *Splenomegaly.* The spleen enlarges early in the acute attack in all sorts of malaria. When a patient has had many attacks, the spleen may be of enormous size and lead to secondary hypersplenism.

4 *Jaundice.* A mild jaundice due to haemolysis may occur in all types of malaria. Severe jaundice only occurs in *P. falciparum* infection, and is due to specific liver involvement.

CLASSICAL STAGES OF FEVER

In a paroxysm of malaria, the patient may notice the following stages:

1 A cold stage (the patient shivers or has a frank rigor; the temperature rises sharply).

2 A hot stage (the patient is flushed, has a rapid full pulse, and a high temperature is sustained for a few hours).

3 A sweating stage (the patient sweats freely, or is even drenched, and the temperature falls rapidly).

These stages are most often recognized in *P. vivax* infection. In rare cases, the patient may be afebrile in the presence of a very severe *P. falciparum* infection.

Hyperpyrexia may complicate malaria, especially in attacks of *P. falciparum*.

PROGRESS OF THE UNTREATED ATTACK

The natural history of untreated malaria differs with each species.

P. falciparum

Following a single exposure to infection, the patient will either die in the acute attack (a common event) or survive with the development of some immunity and residual anaemia. Attacks may recur over the course of the next year (a phenomenon called recrudescence, due to the persistence of blood forms in small numbers between attacks) but then die out spontaneously in the absence of reinfection.

P. malariae

Following a single exposure to infection, and an incubation period that may extend to many weeks, the patient develops a recurrent fever which occurs at increasing intervals. There may be considerable anaemia, and enlargement of the liver and spleen. If no treatment is given to clear the blood forms of the parasite, recrudescences may occur from time to time for more than 30 years. The severity of the attacks tends to diminish as time goes by, until bouts of fever last only a few days.

P. vivax and P. ovale

P. vivax and P. ovale malaria cause very similar illnesses, with bouts of fever which relapse periodically but irregularly over a period of up to 5 years. These are true relapses and not simple recrudescences, because they may occur despite treatment with drugs that entirely eliminate the parasites from the blood. The relapses are due to reinvasion of the blood by merozoites produced when hypnozoites awake from dormancy and develop into PE schizonts.

PECULIARITIES OF P. FALCIPARUM INFECTION

The important difference between P. falciparum and the other plasmodia that infect humans is the capacity of P. falciparum to cause severe (or complicated) disease. Nearly all of the million or more malaria deaths that occur each year result from P. falciparum infections. In endemic areas, most of the clinical impact of P. falciparum infections falls on young children. Nevertheless, the majority of infections cause only a self-limiting febrile illness or, as immunity increases, no illness at all. For reasons that are still not understood, some infections progress to severe disease, and some of these are fatal. In areas with limited or unstable trans-mission, adults (including tourists) with P. falciparum infection may develop severe or complicated disease, especially if diagnosis is neglected or delayed.

MICROCIRCULATORY ARREST

Organs most affected	Main symptoms or signs	Typical misdiagnosis
Stomach and intestines	Vomiting and diarrhoea	Gastric flu; cholera (diarrhoea is *not* bloody)
Brain	Delirium Disorientation Stupor Coma Convulsions Focal neurological signs	Encephalitis, meningoencephalitis (there may be misleading CSF abnormalities)
Kidneys	Renal failure, with or without oliguria or haemoglobinuria	Nephritis
Liver	Jaundice and fever	Hepatitis
Lungs	Pulmonary oedema	Pneumonia, heart failure

Table 1.1 Microcirculatory arrest in P. *falciparum* infection.

Complicated *P. falciparum* malaria may take a number of clinical forms, which are listed in Table 1.1. In young children in endemic areas, who suffer the greatest malaria mortality, four clinical syndromes predominate: these are severe anaemia, cerebral malaria, acidosis and hypoglycaemia. A child may suffer from just one of these complications, or from any combination of them. Other complications seen in adults are unusual in children in endemic areas. Non-immune adults may develop any combination of the syndromes listed in Table 1.1.

Cerebral malaria

This is the most important lethal complication of *P. falciparum* malaria, and occurs only in 'non-immunes'. Its invariable characteristic is a diffuse disturbance of cerebral function, the first manifestations usually being disturbance of consciousness or fits. Focal neurological signs sometimes occur. Untreated, it usually progresses to coma, brain stem failure and death. If recovery does occur, a minority of patients (5–10%) are left with a neurological deficit. Clinically obvious sequelae may resolve over a period of months, but some are permanent. We do not know how many individuals may suffer more subtle impairment (e.g. of memory or intelligence) after cerebral malaria.

The pathogenesis remains unclear. The usual histopathological finding in fatal cases is the presence of large numbers of erythrocytes containing mature parasites in the capillaries and venules of many organs, including the brain. Since irreversible brain damage is unusual in those who recover, it seems unlikely that the microcirculation is totally obstructed by these sequestered cells. The highly active, developing and dividing parasites may consume essential nutrients such as oxygen and glucose and release toxic products, including lactate, with detriment to surrounding tissues. As the schizont ruptures the red cell, substances are released which are known to stimulate the release of cytokines from host cells: these in excessive local concentrations may contribute to coma and other complications of falciparum malaria.

Cases with perivascular 'ring haemorrhages' found at autopsy probably represent an irreversible stage of the disease.

Raised intracranial pressure is usual in children with cerebral malaria, but the mechanism for this is not known, and there is no firm evidence that the raised pressure itself contributes to mortality. Cerebral oedema does not appear to be an important feature of cerebral malaria, at least in adults.

A minor degree of disseminated intravascular coagulation (DIC) is common in falciparum malaria, and DIC severe enough to cause bleeding is an occasional complication in adults.

Hypoglycaemia

Hypoglycaemia is a common complication of untreated falciparum malaria in children. It may occur in adults—pregnant women are particularly susceptible. Hypoglycaemia may also develop as a complication of quinine or quinidine therapy, probably because these drugs stimulate the pancreas to secrete insulin.

Blackwater fever

This obsolete term used to be applied to the syndrome which sometimes occurs in *P. falciparum* malaria when severe intravascular haemolysis is associated with haemoglobinuria and renal failure. The syndrome still occurs, especially in non-immune adults with severe *P. falciparum* infection. In children in the endemic areas of sub-Saharan Africa, haemoglobinuria sometimes occurs in falciparum malaria, but it is rarely accompanied by renal failure. In some cases haemoglobinuria is precipitated by a drug or dietary factor in an individual with glucose-6-phosphate dehydrogenase (G6PD) deficiency. Haemolysis in this condition usually only affects the older cells, so ceases when the haemoglobin has dropped to about 6 g/dl.

MALARIA IN PREGNANCY

All types of malarial infection can lead to abortion. In *P. falciparum* infection, even in women normally immune, pregnancy is associated with an increased likelihood of developing parasitaemia and with higher parasite densities, especially in the first pregnancy. Anaemia is a common consequence, and many women enter labour with a dangerously low haemoglobin. Organ complications such as coma and renal failure are rare in pregnant women living in endemic areas, but among the non-immune, pregnant women are liable to the same complications as other adults.

P. falciparum in endemic areas is an important cause of low birth weight, especially in first-pregnancy babies, who are then at increased risk of dying in infancy from any of a variety of causes. Low birth weight due to maternal malaria presumably results from the fact that the placenta becomes packed with late-stage parasites, especially in the first pregnancy.

In endemic areas it is common to find malaria parasites in umbilical venous blood; it is less common to find them in the neonate's peripheral blood, and these usually disappear within the first 2 days of life. Illness due to congenital infection is rare in endemic areas, but may develop in infants born to non-immune mothers. *P. vivax* is a more common cause of congenital malaria than *P. falciparum*: the illness presents within a few days or weeks of birth with fever, haemolytic anaemia and failure to thrive.

IMMUNITY IN MALARIA

Immunity in malaria is most pronounced in *P. falciparum* infection. In areas of very high transmission, if a child survives to the age of 5 or 6 years, he or she is likely to have achieved a high degree of immunity to the lethal effects of the infection. This immunity has two main components: an ability to limit parasitaemia by the development of specific protective immunoglobulin (IgG) and cell-mediated immunity (antiparasitic immunity), and a physiological tolerance such that low parasitaemia produces no fever or subjective illness (antitoxic immunity). In order to maintain this immunity, frequent re-exposure to infection is required. If re-exposure does not occur, the immunity wanes over a period of a few years. Although West African students living in the UK gradually lose their protective antibodies over a 5-year period, they rapidly regain immunity on re-exposure to infection, but the price may well be two or more severe attacks of malaria on first returning home. The development of a high degree of immunity in an entire population exposed to high levels of

P. falciparum infection has an extremely important effect on the epidemiology of the infection.

Naturally acquired immunity may be suppressed not only by pregnancy but also by major surgery, severe illness of any type and immunosuppressive drugs. However, most studies so far have failed to show increased susceptibility to malaria infection or disease in persons with human immunodeficiency virus (HIV) infection or even in those with acquired immunodeficiency syndrome (AIDS).

Non-immune protective factors in malaria

There are several non-immune factors which affect susceptibility to malaria. *P. vivax* is unable to infect red cells lacking the Duffy blood-group antigen. This is believed to account for the natural resistance of those of pure Negro race to infection with this parasite.

Individuals whose haemoglobin genotype is AS are resistant to the lethal effects of *P. falciparum* infection, but no more resistant to infection itself than those with normal (AA) haemoglobin. This is because the sickle trait prevents the development of high parasitaemia, probably partly as a result of parasitized red cells sickling in the circulation and being removed by the spleen before they can develop into schizonts.

Evidence that G6PD deficiency has a similar protective effect does exist, but is less striking.

Sickle-cell anaemia itself is not protective, for malarial infection is disastrous in such patients.

There is now good evidence that the β-thalassaemia trait confers protection against *P. falciparum*. There is also strong circumstantial evidence that malnutrition protects against the lethal effects of *P. falciparum* infection.

Blood changes in malaria

During the malarial attack the haemoglobin level tends to fall. At the same time the total white cell count also falls, as do the platelets. This fall in platelet count, especially dramatic during an attack of *P. falciparum* malaria, may be useful in deciding if, in a holoendemic area, parasitaemia and symptoms are causally related. The overall level of globulin in the blood rises as a result of repeated attacks of malaria. Some of this globulin only is specifically protective.

Immune disorders in malaria

Some complications of malaria are related to immune effects. Malarial nephrosis is a nephrotic syndrome in children associated with *P. malariae* infection. Antigen–antibody complex is bound firmly to the glomerular basement membrane. An intractable nephrotic syndrome results, with non-selective proteinuria and a bad prognosis. Neither treatment with corticosteroids nor eradication of the malaria seems to influence the outcome. The intractability of the condition seems to be determined by the permanence of the complex-binding mechanism.

A more tractable condition is hyperreactive malarial splenomegaly (formerly known as tropical splenomegaly syndrome), in which marked splenomegaly in *P. falciparum* infection is associated with infiltration of the hepatic sinusoids with lymphocytes, with or without features of secondary hypersplenism. Serum IgM levels are very high. This condition usually resolves in a few months if the patient is given continuous effective chemoprophylaxis.

There is good evidence that an acute attack of malaria has general immunosuppressive effects. The effects of chronic malaria are less well-defined. Interventions against malaria (e.g. bed-nets) have sometimes led to a fall in mortality from other causes, including respiratory infections, suggesting that malarial immunosuppression may increase susceptibility to other common pathogens.

The immunosuppressive effects of malaria may account for the tendency of the Epstein–Barr virus to produce Burkitt's lymphoma in malaria-endemic areas.

DIAGNOSIS

Direct diagnosis

The specific diagnosis of malaria is made by examining the blood, by making a film, drying and staining it. A Romanowsky stain is used, so that the cytoplasm of the parasites stains blue and their nuclear chromatin red. The optimum pH for staining is 7.2, which usually requires the use of buffered water as a diluent. The main stains used are:

1 Field's stain: a rapid method used for staining single thick films only.

2 Leishman's stain: a fixing/staining method suitable for staining single thin films.

3 Giemsa's stain: for staining thin films (after fixation with methanol) and thick films. It is particularly suitable when many films have to be stained together, such as in survey work.

Thick versus thin films

The thin blood film shows the undistorted parasites within the red cells. It is of most use in the detailed study of parasite morphology and species identification. Its disadvantage is that it requires a very prolonged search to detect a low parasitaemia, so its sensitivity is low. A patient may have a fever due to *P. falciparum* and yet have no parasites detected by searching the thin film for half an hour.

The thick film, in which cells are piled upon each other 10–20 deep and lysed and stained at the same time, allows far more red cells to be examined at a time, but it has the disadvantage that the parasites in the lysed cells are distorted. Although readily recognizable as malaria parasites, their specific features of identification may be ambiguous or entirely lost. But the thick film is, in experienced hands, the best method to use for answering the question, 'does the patient have malaria?' The experienced microscopist will often be able to make a specific diagnosis on the thick film alone, despite its shortcomings. Only in difficult cases will uncertainty have to be resolved by examining the thin film as well.

Reservations about blood films

1 One cannot *exclude* malaria by a single negative blood film, for which at least three, taken at intervals, must be examined.
2 A positive blood film does not prove the patient is suffering from malaria (see the case histories, p. 36), because parasitaemia may be entirely asymptomatic in the indigenous population of endemic areas.
3 Some parasites may be washed off the blood films in the staining process (usually only significant with gametocytes of *P. falciparum* in thick films).
4 When slides are stained in bulk, parasites washed off one slide may occasionally find their way on to another.
5 Although there is a statistically greater risk of severe disease in individuals with heavy parasitaemia than in those with few parasites in the blood, the correlation between density of parasitaemia and disease severity is weak. One patient may be critically ill with a low or even undetectable parasitaemia, while another may be relatively or completely well despite having a high parasite count.

Serodiagnosis

Serodiagnosis of malaria is of no use for diagnosis of the acute attack. It depends on finding specific antibodies, and most methods in common use are incapable of distinguishing between antibodies to the different species of parasite. Antibodies may be detectable for several years after the last attack of malaria. The main use of serodiagnosis is in excluding malaria in a patient suffering from recurrent bouts of fever who does not present during a bout. This problem is most commonly met in old soldiers. Serology may also be used in surveys as an approximate measure of exposure of a population to malaria. The most frequently used serological technique is the indirect fluorescent antibody test (IFAT).

Malaria pigment

Malaria pigment may be useful in retrospective diagnosis. In a patient who has recently recovered from an acute attack, pigment may sometimes be found in monocytes for several weeks. These can easily be recognized in a conventional blood film. In patients dying from malaria, the liver, brain and spleen commonly contain large amounts of pigment, which gives the organs a typical dark colour. In haematoxylin and eosin-stained sections, capillaries that are filled with *P. falciparum* schizonts only show the clumps of dark-brown pigment that mark the centre of each schizont. The schizonts themselves do not stain unless impression smears taken from the organs are stained specially with a Romanowsky stain such as Giemsa.

New methods of diagnosis

Many new techniques for identifying malaria parasites are being developed. The quantitative buffy coat (QBC) technique makes use of the fact that parasitized erythrocytes have a different specific gravity from unparasitized red cells and can therefore be looked for in a particular segment of the blood in a centrifuged capillary tube. A dipstick method is now being marketed, by which parasite antigens are detected by placing a drop of blood on a dipstick impregnated with antibody. Polymerase chain reaction (PCR) can be used to detect parasite DNA. At present these are research tools, but their incorporation into clinical practice is probably imminent.

TREATMENT: GENERAL

The treatment of a patient with malaria is supportive and specific.

Supportive treatment

Supportive treatment may include:

1 Reducing the temperature if hyperpyrexia is present – especially common with *P. falciparum* infection.

2 Rehydration, especially when vomiting and diarrhoea have been prominent. Overhydration must be carefully avoided, by weighing the patient if possible.

3 Monitoring renal output and taking corrective measures if necessary.

4 Monitoring the haemoglobin: blood tranfusion, which is sometimes life-saving, should only be given when there are strong clinical indications. In most patients the haemoglobin rises rapidly when the attack has been terminated by specific chemotherapy.

5 Terminating convulsions with appropriate drugs.

6 Monitoring of blood glucose and correction of hypoglycaemia where necessary.

7 Treating DIC if this complication is severe enough to cause bleeding; fresh whole blood, platelet-rich plasma and fresh frozen plasma may be given according to availability.

8 Reducing acidaemia: usually rehydration and antimalarial therapy are sufficient for this purpose. The use of bicarbonate infusion is not of proven benefit, but may be attempted with care in severe acidosis.

Specific chemotherapy

Specific treatment is directed to terminating the parasitaemia as rapidly as possible. The drug of choice depends on national policy in the country where you work, and on the likely place of origin of the patient's parasites. Drug resistance is an increasing problem throughout the world, and the picture changes with time; in some countries multidrug resistance threatens to make malaria untreatable, and new additions to the armamentarium of drugs are urgently needed. In general, treat non-severe malaria with oral drugs if the patient can take them. Complicated falciparum malaria requires parenteral antimalarial drugs, at least until there is clinical improvement and the patient can swallow.

Drugs that prevent the development of the blood stages which are causing the illness are traditionally called schizonticides. Some of them also act against the gametocytes of some species, but this has no relevance to the clinical situation. Some of the schizonticides have useful anti-inflammatory effects also.

The most widely used schizonticide has until recently been chloroquine, but the spread of parasite chloroquine resistance has limited the use of this drug in recent years. However, chloroquine remains the first-line treatment for non-severe falciparum malaria in some semi-immune populations in Africa, and it is the drug of choice for all non-falciparum malarias, although early reports of *P. vivax* resistance to chloroquine are appearing.

Chloroquine

This is a synthetic compound of the 4-aminoquinoline group. It is a bitter white powder. As a base, it forms salts with acids. Those in common use are: diphosphate (Aralen, Resochin, Avloclor) and sulphate (Nivaquine). A slight modification makes hydroxychloroquine; the sulphate (Plaquenil) is used mainly in the treatment of rheumatoid arthritis but occasionally for malaria.

CHLOROQUINE PREPARATIONS

Tablet diphosphate 250 mg (containing 150 mg base).
Tablet sulphate 200 mg (containing 150 mg base).
Tablet sulphate 100 mg (containing 75 mg base).
Coated pill diphosphate 125 mg containing 75 mg base (for children).
Tablet hydroxychloroquine sulphate 400 mg (containing 300 mg base).
Syrup sulphate containing 10 mg base in 1 ml (for young children).
Solution of sulphate for injection 40 mg base/ml.
Solution of diphosphate for injection 50–100 mg base/ml.

ACTION

It is a powerful schizonticide; it also has anti-inflammatory action.

TOXICITY

It is relatively non-toxic if an overdosage is avoided and intravenous infusion is not too rapid. Even moderate overdosage in children may be fatal if the drug is given parenterally. The drug is taken up by the liver, so higher blood levels follow parenteral rather than oral administration. For this reason, the single dose by injection is less than the maximum dose by mouth, although the total dose given is the same in each case.

MAIN TOXIC EFFECTS

Gastrointestinal (nausea, vomiting, etc.); a fall in blood pressure, generalized itching (a common complaint in black-skinned people only); the hair may turn white with chronic overdosage; the vision may also be affected with prolonged use (acute effects–corneal crystal deposition: chronic effects–retinopathy).

DOSAGE

All doses are expressed as dose of base as this is the active part of the drug.

For adults

Oral: 600 mg initially, 300 mg 6 h later, then 300 mg daily for 2 days (total dose 1.5 g).

Intravenous: 5 mg/kg (maximum 300 mg) infused over 3 h in saline, repeated every 8 h to total of 25 mg/kg.

Intramuscular or subcutaneous: 5 mg/kg (maximum 300 mg) every 8 h to total 25 mg/kg.

When giving chloroquine parenterally, change to oral treatment as soon as the patient can take it.

For children

Oral: The dose should be in proportion to the body weight (using the full dose at 60 kg).

Amodiaquine

This is a 4-aminoquinoline with a molecule which has some resemblance to chloroquine and some to quinine. Preparations include tablets of amodiaquine hydrochloride (Camoquin (PD)) containing 200 mg of base. The dose is the same (in terms of base) as for chloroquine. Toxic effects are also similar to chloroquine, but agranulocytosis has been reported.

Amodiaquine is effective against some strains of chloroquine-resistant *P. falciparum* both *in vivo* and *in vitro*; it is no longer recommended for prophylaxis because of the high incidence of agranulocytosis when used for this purpose.

Quinine

Quinine is just as powerful as chloroquine—some think slightly more so. Its main use is in areas of chloroquine resistance. Its isomer, quinidine, is even more active. It is a natural alkaloid derived from cinchona bark, a bitter crystalline powder, practically insoluble in water. It forms salts of varying solubility:

Quinine sulphate (1:500).

Quinine hydrochloride (1:16).

Quinine bisulphate (1:10).

Quinine dihydrochloride (1:0.6).

PREPARATIONS

The content of the preparations is expressed as salt, not base:

Tablet bisulphate 300 mg.

Tablet sulphate 300 mg (also available in capsules).

Injection hydrochloride.

Injection dihydrochloride (usually 300 mg/ml).

Hydrochloride is best for im injection in children, as it is the least likely to cause necrosis.

ACTION

It is a powerful schizonticide; it also has an anti-inflammatory action.

TOXICITY

Tinnitus, deafness, dizziness, nausea, vomiting (this complex of symptoms, known as cinchonism, is almost inevitable with normal doses of quinine; therapy does not need to be stopped or changed on account of such symptoms unless they are severe); hypoglycaemia (especially in pregnancy); hypotension (if excessive dose or if given too fast intravenously); thrombocytopenia (a rare idiosyncratic reaction); erythematous rash. Overdose of quinine can cause deafness, blindess, and severe hypotension. Quinine has a stimulatory effect on uterine muscle, and overdose may cause abortion. (This is not a reason to avoid quinine in pregnancy, since the benefit of curing malaria greatly outweighs the risk of uterine excitation from therapeutic doses of the drug.)

DOSAGE

All doses are expressed as dose of salt.

For adults

Oral: 600 mg 8-hourly for 7–14 days (usually as sulphate).

Parenteral: The intravenous route is preferred. First (loading) dose 20 mg/kg (maximum dose 1400 mg) of quinine dihydrochloride infused over 4 h in an isotonic glucose–electrolyte fluid (e.g. half-strength Darrow's–5% dextrose); subsequent doses 10 mg/kg similarly infused over 2 h, at 12-hourly intervals. Change to oral therapy as soon as the patient can take it.

Intramuscular injection of quinine dihydrochloride may be given if intravenous infusion is impossible. It is usually well tolerated if given deep, with aseptic precautions. Usual dose 10 mg/kg, every 8 h until oral therapy can be substituted.

Chloroquine resistance

Treatment of chloroquine-resistant *P. falciparum* infections: use quinine as already mentioned, together with combination therapy with sulfadoxine–pyrimethamine (Fansidar) or tetracycline to prevent recrudescence in case of RI resistance (p. 24) to quinine.

Fansidar tablets contain sulfadoxine 500 mg and pyrimethamine 25 mg. There is also an injectable preparation containing 500 mg sulfadoxine and 25 mg pyrimethamine in 2.5 ml ampoules for intramuscular injection. The dose for adults is three tablets as a single dose. For children the dose is 0.5–2 tablets according to body weight.

Mefloquine

This is a 4-quinoline methanol drug developed by the US Army and is chemically related to quinine. It is effective against most multidrug-resistant strains of P. falciparum. It is bound to plasma, has a half-life of 21 days, and is effective as a single adult oral dose of 750–1000 mg. It has the big disadvantage that no parenteral preparation is available, and naturally occurring RI resistance has been reported. It is available in 250 mg tablets as Lariam. It is also available in a combination tablet called Fansimef, each tablet containing 250 mg mefloquine, 500 mg sulfadoxine and 25 mg pyrimethamine. Minor toxic effects include headache, dizziness and disturbances of sleep. Occasional severe toxic effects are fits, psychomotor disturbances and psychoses. Individuals with a history of convulsions or neuropsychatric disease are therefore advised not to use mefloquine. There have been many hundreds of well-observed cases in which the drug has been used in pregnancy, without adverse effect on mother or fetus; nevertheless it is wise if possible to avoid the use of mefloquine in pregnancy on general grounds. Patients on cardiosuppressant drugs or β-blockers should not take mefloquine because of its additional effects on the myocardium.

Halofantrine

This is a phenanthrene methanol compound marketed as Halfan which is also active against multidrug-resistant P. falciparum. Despite some misgivings about irregular absorption, three doses of 500 mg at 6-hourly intervals have been highly effective in adults. Its half-life is 1–2 days. Tablets are 250 mg and there is a suspension for use in children.

Halofantrine in high doses consistently causes prolongation of the QTc interval of the electrocardiogram. In occasional individuals there has been possible serious cardiac toxicity. Halofantrine should therefore not be given as treatment to a person who has used mefloquine for current or recent prophylaxis, or for those taking other drugs that suppress myocardial function or conductivity.

Qinghaosu

This is a naturally occurring sesquiterpene lactone originating in China, which is derived from the plant *Artemesia annua*. Its active components, arteether, artemether and artesunate appear to be highly effective antimalarials, active against chloroquine-resistant and multidrug-resistant *P. falciparum* and useful in the treatment of severe and complicated malaria as well as uncomplicated disease. Parasites are cleared from the circulation faster by qinghaosu derivatives than by quinine or chloroquine, but whether this is accompanied by reduced mortality in severe disease remains to be determined. The best way to deploy these drugs in malaria treatment and control programmes is now being carefully studied.

DRUG RESISTANCE IN GENERAL

The problem of drug resistance is most important in *P. falciparum*. There is widespread but patchy resistance to proguanil and pyrimethamine wherever *P. falciparum* occurs. Chloroquine resistance has been known for years to occur in parts of South-east Asia (mainly Thailand, Malaysia and the Philippines), South America and Oceania. In the past decade the prevalence and degree of chloroquine resistance have increased greatly in sub-Saharan Africa, at first in the east and later in the west of the continent. Because resistance is not total, chloroquine remains a useful first-line therapy in some countries, but in others the drug is no longer used, even for uncomplicated disease.

Classification of chloroquine resistance

Chloroquine resistance is classified into three types, according to the reponse of the parasite to normal full doses of chloroquine in patients who are non-immune.

RI RESISTANCE

The patient's clinical response to treatment is normal (satisfactory) and the parasitaemia is apparently terminated. But within 4 weeks parasites are again found in the blood. The assumption is that some parasites do persist in the blood, but in such small numbers as to escape detection. RI resistance can be subdivided into two types, according to whether recrudescence is early or late.

RII RESISTANCE

There is a greater degree of resistance than RI. The patient improves when chloroquine is given, but parasites, although diminishing in number during treatment, can nevertheless be found throughout the treatment period. As soon as treatment ceases, the parasitaemia increases and the patient's clinical condition deteriorates.

RIII RESISTANCE

This implies complete resistance, for neither the clinical condition nor the parasitaemia improves on treatment. Typically the patient goes downhill and the parasitaemia increases relentlessly unless chloroquine is replaced by an effective treatment.

Wherever chloroquine resistance occurs, the RIII resistance pattern is usually less common than the RI or RII patterns. It is obviously impossible to detect the earliest stages of chloroquine resistance in the immune indigenous population, for in these people the effectiveness of chemotherapy is greatly enhanced by the humoral immunity conferred by antimalarial antibodies. The visiting tourist who acquires malaria is an important indicator of drug sensitivity, for the infected visitor can usually be assumed to possess no significant immunity.

Recently a number of cases of *P. vivax* malaria originating in South-east Asia have proved to be unresponsive to chloroquine. Since vivax malaria is not potentially fatal, failure of initial treatment is not a catastrophe, and alternative drugs may be used for the patient who is not cured by chloroquine.

THE PROBLEM OF RELAPSE

Relapse in *P. vivax* or *P. ovale* malaria (see p. 11) can usually be prevented by giving a course of primaquine, but some strains of *P. vivax* (e.g. from Papua New Guinea) are resistant to normal doses of the drug.

Primaquine

This is a bitter, white powder, a synthetic drug of the 8-aminoquinoline group.

PREPARATIONS

Tablets primaquine diphosphate 26.5 mg containing 15 mg primaquine base (or tablets half this size). The dosage is usually expressed as the weight of the base.

ACTION

It is a weak schizonticide. There is action on hypnozoite forms of *P. vivax* and *P. ovale*, and it destroys gametocytes of all species.

TOXICITY

Gastrointestinal disturbance, especially abdominal cramps. Acute haemolysis in G6PD deficiency. Methaemoglobinaemia (cyanosis) usually only with high doses.

DOSAGE

For radical cure of relapsing forms of malaria: 15 mg base daily for 10–14 days. The dose may have to be doubled in some vivax strains. A single weekly dose of 45 mg base for 6 weeks is better tolerated by G6PD-deficient patients than is daily dosing with the lower dose. For clearing gametocytes of *P. falciparum*: 15 mg daily for 5 days. (This use of primaquine does not benefit the individual: it has been advocated in order to reduce the transmission of *P. falciparum*. In highly endemic areas this is a waste of time, since most transmission occurs from individuals who are not even known to carry the parasite.)

CHEMOPROPHYLAXIS

Chemoprophylaxis of malaria involves the regular administration of drugs to prevent clinical symptoms. Drugs taken this way act in two ways: as schizonticides, so that when the parasites enter the red cells they are destroyed, and causal prophylactics. Causal prophylactics prevent the development of the PE schizonts in the liver, and they may also have blood schizonticidal effects. The practical importance of these two modes of action has only become apparent fairly recently, when it was discovered that some drugs may have a much greater effectiveness when given as causal prophylactics, i.e. before sporozoite challenge rather than afterwards.

Drug resistance is tested in two main ways: *in vivo*, when the drug's effect on parasitaemia is measured by daily blood films in carefully supervised patients, and *in vitro*. In the *in vitro* method, parasitaemic blood is incubated with the antimalarial drug at different concentrations, and the proportion of trophozoites that develop into schizonts is measured. In this way, a minimum inhibitory concentration (MIC) can be established.

But neither of these methods gives any indication of causal prophylactic effect. There seems no doubt that even when a certain drug proves itself an ineffective schizonticide by both these tests, it may nevertheless provide a high level of protection if taken before sporozoite challenge. Unfortunately, *in vivo* tests of causal prophylactic effect are difficult and time-consuming and *in vitro* tests do not yet exist.

The upshot of this is that, just because a drug fails on the standard *in vivo* and *in vitro* tests currently in use, it does not necessarily mean that the drug will not protect against malarial attack.

In considering malaria prophylaxis, one is always concerned with balancing risk and benefit. It takes a long time for the real incidence of the toxic effects of a new drug to emerge, and just as long to find out how rapidly resistance develops to it.

One must always remember that chemoprophylaxis never provides *complete* protection against malaria. Everyone embarking on it should be aware of this.

Proguanil (Paludrine, Chlorguanide)

This is a synthetic biguanide. It is a bitter white powder available as hydrochloride.

PREPARATION

Tablet proguanil hydrochloride 100 mg; tablet proguanil hydrochloride 25 mg (paediatric).

ACTION

It is a slowly acting schizonticide and a causal prophylactic. When a mosquito takes up gametocytes from a patient receiving proguanil, their development in the mosquito is inhibited, so the mosquito fails to become infective.

TOXICITY

Proguanil is the safest of all antimalarials: no deaths have ever been recorded from overdose (up to 14.5 g). Occasionally it causes heartburn or epigastric pain, but this is minimized by taking the drug after food. Mouth ulcers have been reported. Gross overdose may cause haematuria. It is safe in pregnancy in a normal dosage, but a folic acid supplement should be given.

DOSAGE

It is used as a prophylactic only. The adult dose is 200 mg/day. It is well tolerated by children who can take 25 mg/day from infancy, 50 mg/day from age 2, 75 mg/day from age 4 and 100 mg/day from age 6 years.

PROGUANIL IN COMBINATION

It has been used successfully in areas of chloroquine resistance in combination with dapsone, in the ratio of two parts proguanil to one part of dapsone, but agranulocytosis has been reported. No combined preparation is marketed.

Pyrimethamine (Daraprim, Malocide)

This is a white, tasteless pyrimidine compound. It is available as the base. It is for oral administration only (but see the section on Fansidar, above, for an injectable preparation combined with sulfadoxine). *P. falciparum* is commonly resistant.

PREPARATIONS

Tablet pyrimethamine 50, 25, 6.25 mg; elixir of pyrimethamine (paediatric) containing 6.25 mg/5 ml.

ACTION

As for proguanil. Its prolonged half-life allows maintenance of effective blood levels with a weekly dosage. It prevents the development of gametocytes in mosquitoes feeding on drug-treated hosts, as does proguanil.

TOXICITY

There are no toxic effects with normal dosage. A dose of 25 mg daily for 6 weeks causes megaloblastic anaemia in adults. It is safe in normal dosage in pregnancy, but a folic acid supplement should be given. It is very toxic in overdose, especially in children. Six adult-sized tablets are fatal in an infant: death is due to effects on the central nervous system (CNS; coma, convulsions). There is no known antidote.

DOSAGE

It is used by itself for suppression only. It is useful for treatment when used in combination (see the section on Fansidar, above and on Maloprim, below). The adult dose is 25 mg/week if the weight is less than 50 kg and 50 mg/week if the weight is over 50 kg. In children 1–3 years: 6.25 mg/week; 4–6 years: 12.5 mg/week; 7–10 years: 18.75 mg/week; over 10 years: adult dosage (according to the body weight as above).

Chloroquine for suppression

DOSAGE

Normal adult dosage is 300 mg base/week. There is virtually no danger in taking this dosage of chloroquine for short periods (say less than 3 years). But as chloroquine binds firmly to melanin (including the pigment of the retina), there is much anxiety that long-term dosage at this level may lead to retinal damage. This has been well documented when chloroquine has been given in high dosage for the treatment of rheumatoid arthritis and related diseases. It is commonly believed that total dosage (because of binding) may be more important in the genesis of retinotoxic effects than the mean daily dosage. This is why many authorities are anxious—and we share their view—about the long-term effects of prophylactic chloroquine. The most widely accepted 'danger level' is a total dose of 100 g of chloroquine base. For this reason we do not recommend chloroquine any more for people who require chemoprophylaxis for more than 6 years on 300 mg base/week continuously. It may be that this view is overcautious, but time alone will tell.

It seems likely that the large-scale use of chloroquine for prophylaxis has hastened the emergence of chloroquine resistance.

Mefloquine prophylaxis

As a result of the spread of chloroquine resistance around the world, mefloquine (alone) is now the prophylactic drug of choice for many areas. Initial anxieties about drug accumulation have diminished, and it is now acceptable to recommend an adult dose of 250 mg weekly for periods of a year or more. Contraindications (see p. 23) must be remembered.

Doxycycline

This long-acting tetracycline is an effective prophylactic against malaria in a dose of 100 mg daily. It is useful in areas where there is resistance to both chloroquine and mefloquine. It should not be used in pregnancy or lactation, nor in young children. An occasional toxic effect is a rash due to photosensitization.

Pyrimethamine–dapsone (Maloprim)

This combination drug has occasionally caused agranulocytosis. Its use in prophylaxis is therefore limited to areas where P. falciparum is resistant to both chloroquine and mefloquine, and where the risk of contracting P. falciparum infection is high. It should be used in combination with chloroquine, because the efficacy of Maloprim against P. vivax is uncertain. The dose of Maloprim should not exceed one tablet per week. (When agranulocytosis has occurred, it has usually been in people taking two tablets per week.)

Prophylactic drugs no longer recommended

Pyrimethamine–sulfadoxine (Fansidar) is not used for prophylaxis be-cause of the risk of Stevens–Johnson syndrome. Amodiaquine should be avoided because of a risk of marrow aplasia.

Caution about chemoprophylaxis

No prophylactic regimen described can be completely depended on to suppress *P. falciparum* malaria, especially in non-immunes. Patients should be warned of this and advised to have an alternative drug available for treatment in the case of failure. Now that resistance has been reported from east and central Africa, it is no longer safe to assume that malaria developing 4 or more weeks after leaving an endemic area will not be *P. falciparum*. It could be that a subclinical parasitaemia, held in check by chloroquine, could increase and lead to a severe attack as soon as the drug is stopped. This is a relatively new danger, but one that must be remembered if tragedies are to be averted.

Nevertheless, if a patient develops malaria more than 4 weeks after leaving a malarious area, it is still most likely that it will be with one of the three non-falciparum species, all of which commonly have a long incubation period.

The effectiveness and safety of the combination drug regimens in children have not yet been fully established.

EPIDEMIOLOGY

The epidemiology of malaria has been most studied in the case of *P. falciparum*. The two most important factors are:
1 Intensity of transmission (the number of infective bites per year).
2 The immune response of the host.

MEASURING MALARIA IN A COMMUNITY

Traditional methods

It has been customary in the past to characterize the epidemiological situation in a community by describing its malariometric indices. These are established by surveys which, by examining all age groups of the population, determine for each group:
1 The parasite rate (the proportion of blood films which are positive).
2 The spleen rate (the proportion of the group with enlargement of the spleen).

Obviously, both these indices may show seasonal variations. In the case of the spleen rate, its reliability will depend on whether or not there are other diseases in the population which can cause enlargement of the spleen, such as schistosomiasis and visceral leishmaniasis. The measurement of spleen size is of much less importance.

Morbidity and mortality

It is now recognized that parasite and spleen rates are measures of malaria infection, reflecting the intensity of transmission, but they are not measures of the clinical impact of malaria on the community. It is the morbidity and mortality attributable to malaria that are important as the basis for designing a malaria control programme, and these indicators are equally important in monitoring the effectiveness of control.

Ways of estimating malaria-attributable mortality include hospital studies and 'verbal autopsies'. The latter technique makes use of tested questionnaires to inquire of mothers about the nature of the final illness in any children dying within a specified period before the survey. Unfortunately, the verbal autopsy technique cannot distinguish reliably between malaria and pneumonia or meningitis, and deaths due to severe malarial anaemia may not be identified. Therefore only an approximate measure of malarial mortality can be obtained.

Malarial morbidity can be assessed by prospective studies of cohorts of people for episodes of fever, by health centre and hospital records of severe disease, and by cross-sectional surveys of a population measuring haemoglobin levels to identify malarial anaemia. The proportion of first babies that are of low birth weight is an indirect measure of the prevalence of malaria in the primigravid women of a community.

Probably the best method of measuring the clinical impact of malaria is to see what happens when an effective intervention against malaria is introduced. In studies of insecticide-impregnated bed-nets it has been shown that malarial morbidity and mortality can, in some areas, be reduced by the intervention, and that the reduction may be accompanied by improved overall survival. This suggests that malaria may contribute to apparently non-malarial mortality, by mechanisms—perhaps including immunosuppression—that are not fully understood.

Stable malaria

Transmission occurs for at least 6 months in the year and is intense. Malarial infection is acquired repeatedly. Children suffer repeated attacks of malaria from the age of a few months onwards (very young children are partly protected by passive immunity acquired by transplacental passage of protective maternal IgG). This may modify the severity of the first few attacks, so allowing them to develop some active immunity while still partly protected. Children reaching the age of 5 or 6 years have substantial immunity, but the price of this immunity is that some children will die of malaria before immunity develops. The proportion who do so is likely to depend on many factors, including the intensity of transmission, the availability of drugs and the prevalence of parasite drug resistance. Data on the actual death toll in different populations are still rarely available.

When immunity has been established, older patients may still suffer attacks of malaria, but these take the form only of mild, flu-like episodes lasting a few days. Severe and complicated disease rarely occurs. In areas of stable malaria, the adult population is little affected. As a result the effects on the working population and the economy are slight.

There is little variation in the incidence of malaria from year to year (hence the word 'stable'), but there may still be pronounced seasonal fluctuations in new cases seen in children. In such an area, there is often a marked rise in the number of children seen with cerebral malaria about 2 weeks after the rains begin.

TRADITIONAL MALARIOMETRIC INDICES IN STABLE MALARIA

The spleen rate is typically high in young children, reflecting the high frequency of infections in this age group. A spleen rate of 75% in children aged 2–6 years is usual, and the rate may be even higher. With the development of immunity the rate falls progressively with increasing age, and the spleen rate in adults is low.

The parasite rate is high at all ages: the rate in children is often 90% or more. In adults the rate remains high (commonly 50% or more) but the density of parasitaemia is low in most adults. This is a reflection of the combination of antiparasitic and antitoxic immunity.

Unstable malaria

The situation is the antithesis of stable malaria. There are wide changes in transmission not only throughout each year, but also from year to year. This results in the tendency for epidemics to occur—hence the name 'unstable'. The transmission season is typically short and the mosquito population fluctuates widely. Infection is usually so infrequent that no member of the population has the opportunity to develop a significant level of immunity. For this reason, when transmission does suddenly increase (usually due to freak environmental conditions leading to an explosion in the mosquito population), people of all ages are equally susceptible to infection. This results in serious disease or even death, regardless of age. The health of the working community may be disastrously affected, and the economic effects of an epidemic can be dire.

We see here a paradox: there are many situations where it is better for the community to have more malaria rather than less. Obviously, the danger of deliberately meddling with stable malaria is that it may become unstable, and so more of a social burden.

TRADITIONAL MALARIOMETRIC INDICES IN UNSTABLE MALARIA

These will vary greatly from year to year and from time to time within the year. During periods of low transmission, parasite and spleen rates will be low in all age groups. During epidemic periods, both indices will be high in all age groups, reflecting the low level of 'herd' immunity.

Varieties of patterns of malarial epidemiology

Professional malariologists recognize the two extreme epidemiological situations described, but also recognize that there are many situations which do not exactly fit into either category. For this reason a variety of environmental patterns (once called paradigms) have been described, which are associated with different epidemiological types of malaria. The number of different environments could be infinite, but eight major categories have been defined. These are:

1 African savannah.
2 Desert and highland fringes.
3 Plains and river valleys outside Africa.
4 Forests.
5 Extensive agricultural developments.
6 Urban areas.
7 Coastal areas and marshlands.
8 War zones and areas of political and social disorder.

The older terms holoendemic and hyperendemic refer to areas with intense year-round transmission of malaria; these are characterized by the stable pattern of malaria. Mesoendemic areas are those with regular transmission confined to a limited season each year, and hypoendemic areas are those with occasional or sporadic transmission.

GLOBAL MALARIA ERADICATION

For many years global malaria eradication was the aim of the World Health Organization (WHO) and was believed to be feasible because of the invention of dichlorodiphenyltrichloroethane (DDT). Based on periodic house-spraying with DDT, and the idea that if the life span of the female anopheline vector could be reduced below 10 days (the minimum extrinsic incubation period), then even if a mosquito had fed on a gametocyte carrier, she would not have time to become infective before her death. Eradication relied on the following assumptions:

1 It would be possible periodically to spray all dwellings in endemic areas.
2 The mosquitoes would rest on sprayed surfaces after taking a blood meal (and so be exposed to the insecticide).
3 The mosquitoes would remain sensitive to the lethal effects of DDT.
4 Individuals would cooperate.
5 Nations would collaborate in both spraying and submitting to treatment with gametocytocidal drugs.

Failure of global eradication

Eradication has only succeeded in a few areas, mostly islands. The main causes of failure have been:
1 *Operational*: not all houses were sprayed. There are many causes for this, including lack of cooperation, poor mapping, accelerated destruction of thatched roofs (DDT kills caterpillars and their predators; caterpillars soon reappear but predators do not) and resentment of intrusion into privacy.
2 *Technical*: resistance of mosquitoes to insecticide; behavioural resistance in which mosquitoes fly straight out of the house after feeding, and so do not rest on the sprayed surface; resistance of the parasite to antimalarial drugs.
3 *Political* (not a cause recognized by the WHO): failure of countries of cooperate; civil war and severe political unrest; lack of a suitable infrastructure on which to build the control programme; political and administrative incompetence.
4 Ill-advised and unpopular pilot schemes.
5 Failure to convince the people of the need for the programme.

Malaria control at present

Global eradication as an objective has been abandoned as unrealistic. Instead the objective now is the control of disease and mortality that result from malaria. This objective requires:
1 The provision of diagnostic and treatment services as close as possible to where the people live.
2 Simple, affordable and safe therapy for uncomplicated disease.
3 Prompt recognition and treatment of severe disease, with systems of referral to hospital centres when necessary.
4 Education of the population in the features and dangers of malaria.
5 Drug prophylaxis for selected subgroups.
6 Antivector measures and water clearance where achievable (and not elsewhere).

7 New interventions (now being assessed), e.g. the use of permethrin-impregnated bed-nets or curtains.

Vaccines may in the near future be added to the list of effective interventions, Patarroyo's synthetic peptide vaccine (SPs66) has been shown to give 30% protection in field trials in Tanzania recently.

Individual precautions

Because the anopheline vectors of malaria are night biters, a high degree of protection is given by:

1 Covering up the exposed skin in the evenings.

2 The use of insect repellents such as dimethyl phthalate, dibutyl phthalate or diethyltoluamide.

3 The use of an efficient mosquito net over the bed, preferably impregnated with a synthetic pyrethroid such as permethrin or deltamethrin.

DRUG PROPHYLAXIS OF MALARIA

This should be used by all visitors to endemic areas.

What to advise for the indigenous population is far more difficult (see Fig. 1.1, pp. 4–5). There is evidence that prophylaxis can reduce the incidence and mortality of malaria in young children, but it is rarely possible for regular prophylaxis of children to be achieved on a large scale. It either costs too much or cannot reliably be delivered.

Current WHO advice is therefore that all children should have prompt access to effective treatment. Because so many children live too far from a medical aid post to avail themselves of treatment promptly, the best solution may be to put the onus of treatment on the mothers, by a health education programme which stresses the importance of prompt treatment of fever. If treatment is delayed, the child may be unable to take oral medication because of vomiting. Such a child will require parenteral treatment, and be at increased risk of progressing to complicated disease because of the delay in obtaining treatment.

Such a policy seems to be the only practical solution to this problem in many tropical countries. Even if attacks are treated effectively, immunity will eventually develop. That chloroquine overdose parenterally is highly lethal is well known. It is not so well known that even oral overdose can kill. The importance of ensuring that mothers are educated to appreciate this is obvious. Alternative first-line drugs (e.g. Fansidar), now being introduced in many chloroquine-resistant areas, carry their own dangers, which in most cases have yet to be defined.

QUESTIONS, PROBLEMS AND CASES

(For answers, see p. 339)

1 How can a person acquire malaria?

2 What is the main reason why *P. falciparum* malaria is sometimes called malignant tertian malaria?

3 In what four ways can malaria cause anaemia?

4 What are the three most important causes of fever with diarrhoea and vomiting in children in tropical areas?

5 What is the most important method for diagnosing acute malaria?

6 The 3-year-old son of an African farmer is brought to your clinic with a 24-h history of fever followed by vomiting and a 1-h history of repeated convulsions. What would you do?

7 An expatriate family with two children aged 2 and 4 years is taking up residence in East Africa for a year. What advice would you give them on malaria prophylaxis?

8 How can you distinguish cerebral malaria from febrile convulsions?

9 A Nigerian medical student is admitted to hospital in the UK with a 1-week history of fever, headache, abdominal pain and constipation. He arrived in Europe 3 days ago from Lagos. On examination he looks toxic, his pulse is 100 beats/min, temperature 39°C, and his spleen is palpable to 3 cm below the left costal margin. The blood film shows scanty, ring-form trophozoites of *P. falciparum*. The white blood cell count (WBC) and differential are normal.
   (a) What is the most likely diagnosis?
   (b) What investigations are likely to be helpful?
   (c) What is the treatment of choice?

10 You are rung up by a 35-year-old British engineer in West Africa. He says he has had a fever for 5 days and has been told by his steward that he had a convulsion 30 min ago. He is now unable to move his left arm or leg. Your examination of the patient is interrupted when he has a further major fit. When this is over, he is found to have left and right hemiplegia (pseudobulbar palsy). His pulse is 120 beats/min and temperature 40°C.

(a) What is the single most important investigation required?
(b) Could this finding explain all his symptoms?
(c) What is the other most important investigation required?
(d) Could the abnormalities in the cerebrospinal fluid (CSF) all be explained by malaria?
(e) How would you manage the patient?
(f) Could intensive corticosteroid treatment help recovery?
(g) The patient's platelet count was found to be low (28 000 µl) and fibrin degradation products were present in the urine. Should heparin be added to his regimen?

Outcome: The patient was treated with intravenous quinine initially (20 mg/kg—total 1200 mg—of quinine dihydrochloride infused over 4 h) and then by oral quinine. No further convulsions occurred after admission to hospital. He made a rapid neurological recovery, and left hospital after 3 weeks with no sequelae.

FURTHER READING

Beadle, C. et al. (1994) Diagnosis of malaria by detection of P. falciparum HRP-2 antigen with a rapid dipstick antigen-capture assay. Lancet **1**, 564–7.

Bradley, D.J. (1993) Prophylaxis against malaria for travellers from the United Kingdom. Report of a meeting convened by the Malaria Reference Laboratory and the Ross Institute. BMJ **306**, 1247–52.

Gilles, H.M., Warrell, D.A. (eds) (1993) Bruce-Chwatt's Essential Malariology, 3rd edn. Edward Arnold, London.

Greenwood, B.M., Bradley, A.H., Greenwood, A.M. et al. (1987) Mortality and morbidity from malaria among children in a rural area of The Gambia, West Africa. Trans R Soc Trop Med Hyg **81**, 478–86.

Hien, T.T., White, N.J. (1993) Qinghaosu. Lancet **341**, 603–8.

Hill, A.V.S. et al. (1992) Common West African HLA antigens are associated with protection from severe malaria. Nature **352**, 595–600.

Molyneux, M.E., Fox, R. (1993) Diagnosis and treatment of malaria in Britain. BMJ **306**, 1175–80.

Warrell, D.A., Molyneux, M.E., Beales, P.F. (eds) (1990) Severe and Complicated Malaria, 2nd edn. World Health Organization Malaria Action Programme. Trans R Soc Trop Med Hyg **84** (suppl 2), 1–65.

Visceral leishmaniasis

Visceral leishmaniasis (kala-azar) is a disease caused by a protozoan complex called *Leishmania donovani* which is widely distributed and comprises three species—*L. infantum* around the Mediterranean basin, through central Asia and China, *L. donovani* in India and eastern Africa and *L. chagasi* in parts of south and central America (Fig. 2.1).

PARASITE

The reservoir of infection is the amastigote form of the parasite, present in animal reservoir hosts such as rodents, dogs, foxes and jackals, or in humans. The amastigote is usually intracellular and is a minute (2–4 μm in diameter), roughly spherical structure containing two distinct pieces of nuclear chromatin. The larger piece is called the nucleus, the smaller piece the kinetoplast.

LIFE CYCLE

The infection is usually transmitted by the bite of female sandflies. These belong to the genus *Phlebotomus* in the Old World and *Lutzomyia* in the New World. The sandfly becomes infected by taking up the amastigotes with its blood meal, the amastigotes being in the blood or skin of the infecting host. The amastigotes are liberated from the cells in which they are usually found, in the stomach of the sandfly, and begin to multiply by simple fission.

The amastigote becomes elongated and fusiform, and a flagellum appears, originating from the kinetoplast which lies anterior to the nucleus. This organism is the promastigote (leptomonad) and is the stage of the parasite which develops when amastigotes from infected animals or patients are grown in culture. The motile promastigotes migrate from the gut to the mouthparts of the sandfly, and are injected during feeding.

Human infection has been reported from blood transfusion and, even more rarely, by direct transmission of amastigotes from person to person in infected secretions, presumably during sexual intercourse. Congenital infection has also been reported.

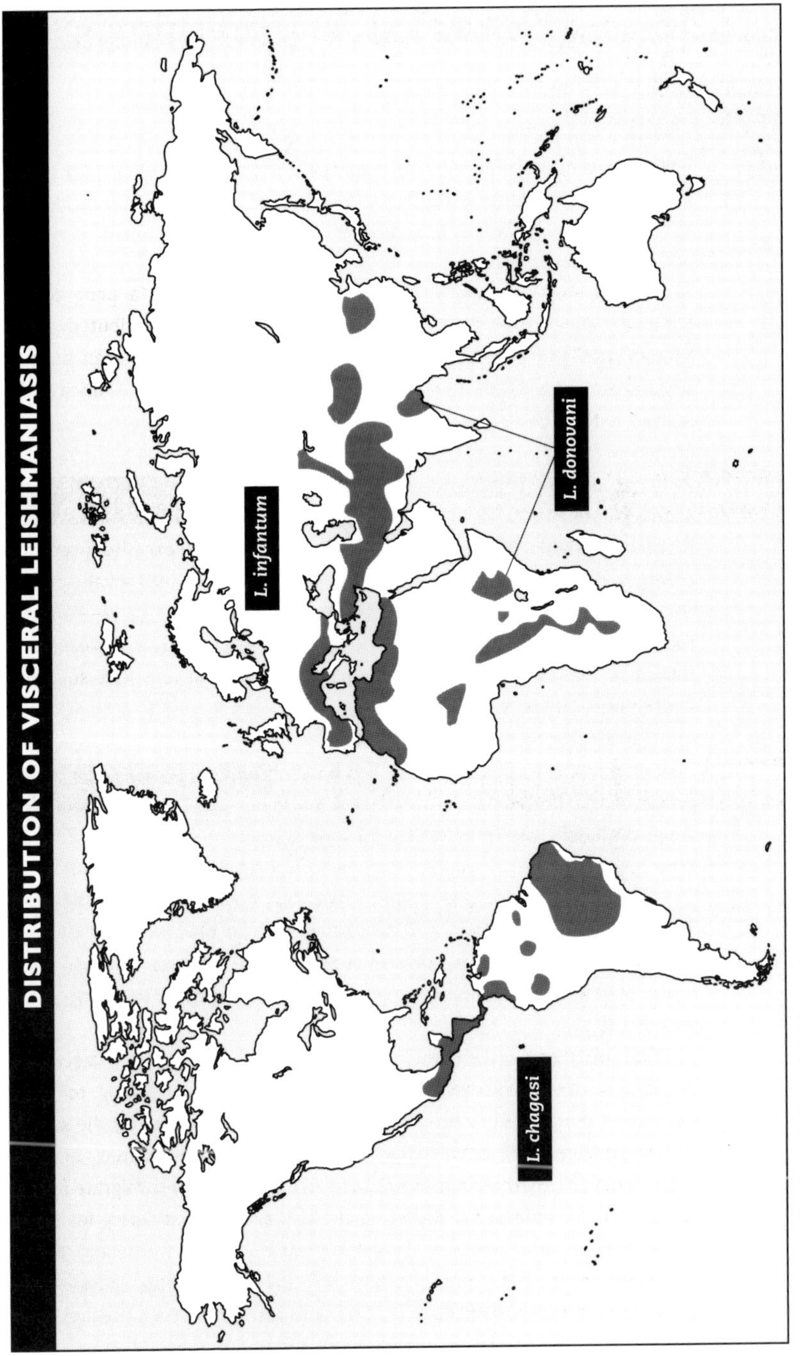

DISTRIBUTION OF VISCERAL LEISHMANIASIS

L. infantum

L. donovani

L. chagasi

Fig. 2.1 Distribution of visceral leishmaniasis.

DISEASE

Initial response

Following the injection of promastigotes, they may multiply locally and excite a cellular reaction leading to the formation of a definite nodule. When such a lesion forms it is called a leishmanioma; its frequency varies with the different geographical strains of leishmaniasis. If the local tissue response is very vigorous, it may lead to death of the parasites and their complete elimination. The cells found in the lesion are macrophages, lymphocytes and plasma cells, but the effective cells in destroying the organisms are probably specialized killer cells. If spontaneous recovery follows natural challenge in this way, the patient's specific cell-mediated immunity (CMI) increases, and the patient will develop a delayed sensitivity response resembling the tuberculin reaction if a suspension of killed promastigotes is injected intradermally. This is the leishmanin or Montenegro test. The development of a positive leishmanin test is accompanied by resistance to reinfection. If the CMI response is poor, the amastigotes multiply rapidly inside the macrophages, eventually leading to rupture of the parasitized cells, and so become disseminated throughout the body. They are taken up by further phagocytic cells of the reticulo-endothelial system, and the visceral stage of the infection is established.

The incubation period is very variable, usually 3–18 months. Occasionally it is as short as 2 weeks. It is now known that the parasites may persist in the body in a dormant state for many years, only to emerge when the host's immunity wanes for any reason.

Established visceral leishmaniasis

The generalized infection leads to fever, usually of slow or subacute onset, although in exceptional cases rigors do occur. Sweating is usually prominent.

The amount of malaise accompanying the fever is usually slight, and the patient often feels relatively well and continues to eat well and to work until the disease is fairly advanced. Typically, the fever is irregular and intermittent at first, but later it becomes remittent. The invasion of phagocytic cells by the amastigotes leads to a great increase in the mass of the reticuloendothelial (RE) tissues, including the spleen, liver and lymph glands.

Enlargement of the spleen (which may be enormous) is usually followed by enlargement of the liver. The enlargement of the lymph glands is a less constant feature, and shows great variation with different geographic strains and in different outbreaks. It is usually most marked in Africa.

By the time the spleen is greatly enlarged, the patient often becomes emaciated. Jaundice due to liver damage occurs in a minority of cases. Pain over the spleen due to small infarcts is common and may be severe, and there are usually gross changes in the splenic capsule at postmortem.

Immune changes

The antigenic stimulus of the countless amastigotes is enormous, and there follows a high increase in the production of immunoglobulin, chiefly IgG. Only some of this is specific (i.e. capable of combining with amastigotes) and even this specifically reactive antibody is lacking in protective effect, perhaps because most of the amastigotes are inaccessible to antibody because of their intracellular habitat.

The diversion of the body's humoral immune resources to manufacturing useless antibody reduces the defences against a variety of bacterial infections, the major cause of death. The specific CMI response is always poor in visceral leishmaniasis, so the leishmanin test is invariably negative during the disease. It often becomes positive after successful treatment.

Typical blood changes

At the same time as the IgG level rises, the total globulin level reflects it, often exceeding 4 or 5 g/dl. The albumin level often falls at the same time, sometimes sufficiently to lead to oedema. Anaemia is a typical feature of visceral leishmaniasis; it is normochromic and normocytic in type. Although initially mainly due to secondary hypersplenism, later there may also be marrow depression as reflected by a relatively low reticulocyte count. There is also a leucopenia, mainly of granulocytes and sometimes amounting to agranulocytosis. Ninety-five per cent of cases when first diagnosed have a total WBC of less than 3000/μl, and in 75% the count is less than 2000/μl. Part of this granulocytopenia is splenic in origin, part due to the effects of the parasites on the marrow. The eosinophil count is invariably low. The platelet count is also usually low, sometimes sufficiently low to lead to purpura, spontaneous bruising and bleeding, especially from the gums, and epistaxis.

Less common localized effects

Involvement of the liver may lead to micronodular cirrhosis, but whether due directly to the parasites or not is undecided. Infection of the nasal mucosa may lead to localized destructive lesions resembling espundia.

Groin glands may be very prominent in the African type of disease and a minority of patients present with a local or generalized lymphadenopathy without other features of visceral leishmaniasis. The small intestinal mucosa may show marked infiltration with parasitized macrophages.

Visceral leishmaniasis and HIV

An increasing number of reports of leishmaniasis associated with HIV infection has been described from southern Europe. In HIV-infected individuals, visceral leishmaniasis presents atypically with skin and respiratory lesions, less frequent splenomegaly, poor leishmanial antibody responses and very poor response to therapy.

NATURAL HISTORY

The untreated disease is usually fatal, but the course of the illness is very variable. A strong granulomatous reaction around the parasites, indicating a high level of CMI, is a good prognostic feature. Visceral leishmaniasis is one of the many causes of non-caseating granulomata in the liver. But in the kala-azar patient, the granulomatous response is minimal.

Death may occur within a few weeks of the first symptoms or be delayed for 2 years or more. A patient in the UK was kept alive by blood transfusions and other supportive treatment for 8 years (having been diagnosed as suffering from aplastic anaemia) before a correct diagnosis was made and her illness cured by specific therapy.

More than 19 out of 20 deaths are caused by intercurrent infection, mainly of the respiratory tract and gut. Pneumonia, bacillary and amoebic dysentery and tuberculosis are all common. Although diarrhoea may be due to a dysenteric infection, there is no doubt that many patients present with diarrhoea due to mucosal involvement by the disease itself.

DIAGNOSIS

There are three main approaches, often used in combination:
1 Direct diagnosis (finding the amastigotes).
2 Indirect diagnosis (finding antibodies).
3 Circumstantial diagnosis (by identifying the usual pathological changes found in the disease).

Only direct diagnosis is incapable of giving false-positive results.

Direct diagnosis

The object is to find the amastigotes and identify them by microscopy, or to culture material containing amastigotes on a medium which allows their growth and replication as promastigotes, or by animal inoculation. Material is obtained from the bone marrow, spleen or lymph nodes. Whilst marrow aspiration is safer, splenic aspiration is more sensitive. Splenic puncture is usually safe if the organ is large and firm but the platelet count should be above 50 000/cm^3 if the risk of bleeding is to be minimal, and the prothrombin time should be checked. An assistant should fix the spleen between two hands and a dry small intramuscular needle (22-gauge) is rapidly inserted into the spleen with suction using a 2 ml syringe without local anaesthetic and immediately withdrawn. The contents should be carefully blown out on to a clean slide and spread into a film.

The same sample can be inoculated on to NNN medium. The film is then dried at ambient temperature, fixed with methanol, and stained with a Romanowsky stain such as Giemsa. The same technique can be used for lymph gland puncture.

When marrow (posterior iliac crest in both adults and children) is examined, suction must be used to aspirate the marrow contents. This procedure is more painful than splenic puncture, and local anaesthesia is always necessary.

Hamster inoculation is largely an academic procedure, as the infection may take 6 months to become manifest. Infections 'coming up' before then kill the animal, whose spleen should be examined by Giemsa-stained impression smear.

In outbreaks of Indian kala-azar, amastigotes may be found in monocytes in the buffy coat of centrifuged blood.

Indirect diagnosis

So-called specific antibodies are sought in the patient's serum. This method can be used for preliminary screening or when marrow or spleen aspiration are negative or puncture and similar direct diagnostic procedures are considered too hazardous because of the risk of haemorrhage. (One of the questions on p. 49 refers to such an instance.)

There are many methods in use, including IFAT and the enzyme-linked immunosorbent assay (ELISA). Although they may be useful, they may show cross-reaction (false-positive results) with a variety of other infections, including malaria, toxoplasmosis and disseminated tuberculosis, although the titres with false-positive reactions are usually lower. Recently an antigen detection test, the direct agglutination test (DAT), has been developed.

Circumstantial diagnosis

This relies on assembling a constellation of separate tests which, together with the clinical features, support the diagnosis. It is the most dangerously misleading method of all, as the following true anecdote illustrates.

In a district in a certain east African country after the Second World War, deaths were reported to be occurring in inhabitants of all ages, following an often prolonged febrile illness. The clinical features were fever, weakness, anaemia and enlargement of the liver and spleen. A medical officer made this outbreak of disease the subject of his special attention. He found the situation as described and, in addition, most of the patients had pancytopenia and hyperglobulinaemia as judged by a positive formol gel test. He wrote up his work under the title 'An outbreak of Pentostam-resistant kala-azar'.

Only when a more experienced medical team moved in and performed blood films in addition to spleen smears was it found that the outbreak of 'visceral leishmaniasis' was, in fact, an outbreak of malaria. That many patients died despite antimonial chemotherapy is scarcely surprising. The point of this story is to illustrate that it is simply not possible, by circumstantial methods alone, to make a definite diagnosis of visceral leishmaniasis. Confusion may arise not only with malaria, but with aleukaemic leukaemia, aplastic anaemia, brucellosis, trypanosomiasis, disseminated tuberculosis and histiocytic medullary reticulosis.

FORMOL GEL TEST

This simple test, when positive, indicates hyperglobulinaemia, whatever its cause. It should never be regarded as specific for visceral leishmaniasis. To 1 ml of serum in a test tube is added one drop of concentrated formalin solution (40% formaldehyde) and the tube is then shaken to mix thoroughly. After 20 min at room temperature, the serum becomes a firm, opaque jelly (like a cooked egg white) if the test is positive.

Post-kala-azar dermal leishmaniasis

This may occur after chemotherapy of visceral leishmaniasis or, less often, after spontaneous recovery. It is a macular or nodular skin eruption, localized or widespread, of very varied appearance. It is caused by the presence of a great number of amastigotes within the skin. It is most common after Indian kala-azar, where it forms the only reservoir of infection. It may respond to long courses of treatment with sodium stibogluconate, or may be completely refractory. In some cases it persists for several years before finally resolving spontaneously.

TREATMENT

Pentavalent antimony drugs

Most cases of kala-azar respond to parenteral pentavalent organo-antimony drugs. Haematological and parasitological follow-up should extend to 1 year, to detect relapse early.

Sodium stibogluconate; sodium antimony gluconate (Pentostam)

This is a pentavalent antimony compound available as a 33% solution equivalent to 10% elemental antimony. It is given intravenously or intramuscularly in a daily dose of 0.1 ml/kg (maximum dose 6 ml) in courses of treatment of varying length. It is much less likely to cause vomiting and other toxic effects than trivalent antimony compounds. Different geographical strains of L. donovani show differences in their susceptibility to antimony; Indian strains are more sensitive than strains from east Africa.

Recent experience suggest that the optimum dose is 10–20 mg/kg each day for a period of 30 days. Unless this dose is given initially, resistance may be induced and relapse occur. Children require a proportionately higher dosage. An equivalent dose of elemental antimony should be used if the other preparations are used.

Other pentavalent antimonials are available in different countries, and seem to be of similar potency, e.g. antimony-N-methyl glutamine (Glucantime) is widely used in south America and urea stibamine (Carbostibamide) is used in India.

The encapsulation of drugs in liposomes shows great experimental promise, enhancing drug effectiveness by delivering it in packets which are taken in by infected macrophages by phagocytosis. Whilst antimonials have not been prepared as commercial preparations as yet, amphotericin is now available as a liposome-encapsulated compound (AmBisome).

Aromatic diamidine drugs

HYDROXYSTILBAMIDINE ISETHIONATE

This is available in ampoules containing 255 or 250 mg dry powder, for reconstitution in water for injection immediately before use intramuscularly or in 200 ml 0.9% saline before administration by intravenous infusion (over 1–2 h). Intramuscular injections are painful.

Dose

3–4.5 mg/kg per day, maximum dose 250 mg daily for 10 days. With 7 days' rest between, three such courses can be given in antimony-resistant cases.

Toxic effects

Mainly hypotension, but this is minimized by slow intravenous infusion. Minor effects include dizziness, nausea, vomiting and pruritus. Antihistamine drugs may reduce the toxic effects.

Aminosidine (paromomycin) has recently been used in the treatment of visceral leishmaniasis – 14–16 mg/kg per day by infusion for 21 days or for 1 week after parasitological cure.

ALLOPURINOL

This has been reported of value in cases unresponsive to Pentostam, at a dose of 16–24 mg/kg given in three divided doses daily for 10 weeks or more.

Refractory cases

The major cause of refractory cases is usually inadequate dosage with antimony, and the physician caring for such a case should be prepared to raise the dose and prolong the course until toxic effects occur before resorting to the more desperate remedies which follow. An electrocardiogram (ECG) will warn of early cardiotoxicity (prolonged QTc interval initially).

Amphotericin B (Fungizone), an antifungal antibiotic, is toxic, and is only used when other treatment has failed.

DOSE

A dose of 250 μg/kg per day initially, increasing to 1 mg/kg per day, given daily by slow intravenous infusion, diluted to 1 litre in 5% dextrose solution, over at least 6 h. A course of treatment normally consists of daily infusions to a total (adult) dose of 2 g, but can be prolonged to several weeks' duration if lack of response shows the need and toxic effects permit.

TOXIC EFFECTS

These are mainly thrombophlebitis and renal damage. Renal function should be monitored throughout treatment, and treatment not restarted

until renal function has recovered. On restarting treatment, begin with
the low dose as when initiating treatment.

SPLENECTOMY

When all else fails, splenectomy may, if combined with drug treatment,
result in cure. Splenectomized patients should receive oral phenoxymethyl penicillin continuously for the first year after operation to protect
them against pneumococcal infections and be given pneumococcal vaccine, although their response to this may be poor. In malarious areas,
they will also need malarial chemoprophylaxis for life.

EPIDEMIOLOGY

Reservoirs and sandflies

The epidemiology varies greatly in different parts of the world because of
the different reservoirs of infection and the different patterns of human
exposure to infection. All sandflies breed in dark, moist habitats, such as
cracks in masonry, in piles of rubble, in caves and in any dark protected
sites such as holes in termite mounds or in outside privies. Sandflies have
a short flight range, being seldom found more than 200 m from their
breeding place. They do not fly very high and seldom bite people sleeping
on the first floor of a building. They normally bite at dawn and dusk, and
also during the night.

   The usual reservoirs of infection are canines in the Mediterranean
basin, the Middle East, central Asia and south America (the relative
importance of domestic dogs and wild canines, such as jackals and foxes,
varies with each area); rodents and humans in Africa south of the Sahara;
and humans in India.

Patterns of human outbreaks

Outstanding features are the tendency of Mediterranean kala-azar to
attack young children much more than adults, the tendency for Indian
kala-azar to develop into large epidemics, and the sporadic and epidemic
nature of infection in east Africa. This is sometimes related to occupational exposure, such as in cowherders using termite mounds as
vantage points to watch their beasts. Sandflies are often found in
such habitats, and the catholic tastes of the sandfly ensure that a hungry
fly will bite a human being as readily as its usual source of food. It is
assumed that under endemic circumstance, rodents are the reservoir of
infection.

Because of the short flight range of the sandfly, outbreaks often affect several members of the family, or are limited to a few closely situated houses. Epidemics in Kenya, Ethiopia and the southern Sudan have occurred, probably by the spread of infection within the human population. In recent epidemics in the southern Sudan a high frequency of post-kala-azar dermal leishmaniasis has been found.

CONTROL

The methods which can be used depend on the epidemiological situation. Most success has been achieved where the domestic dog is the main reservoir, efforts being directed to catching and destroying infected dogs. Dogs with the infection look sick, lose their hair and have an enlarged spleen. France has been very active in control by this method. But infected foxes invariably look healthy and are more difficult to control. Where human beings are the main reservoir, such as in India, successful control was achieved during the period when widespread insecticide spraying of houses for malaria control was in use. The resurgence of infection on an epidemic scale has occurred in several areas many years after the spraying programme was abandoned. In all epidemics and localized outbreaks, residual insecticide spraying of houses and the immediate area around the house is the most effective immediate measure. Sandflies are not resistant to DDT, so their hopping flight pattern renders them particularly vulnerable.

Detection of cases and prompt treatment are important, for blood itself may be infective to sandflies during the active disease. The treatment of post-kala-azar dermal leishmaniasis is also a useful control measure, but its relative resistance to treatment must be remembered.

In the sporadic kala-azar of Africa, where rodents are the main reservoir, there are no practical control measures available. Individual protection against sandfly bite is difficult, because repellents have only a brief effect and sandflies can pass through the mesh of normal mosquito nets. Mesh impregnated with repellent or synthetic pyrethroids can prevent the ingress of sandflies for 2 or more weeks, and good protection against night biting is given by sleeping on the first floor of the house or on the roof.

Although a vaccine is not yet available commercially, the degree of cross-resistance between various *Leishmania* species makes such a development conceivable for the future. Recovery from a natural attack usually leads to lifelong immunity to attack by the same parasite.

QUESTIONS, PROBLEMS AND CASES

1 What are the most important vertebrate hosts of *L. donovani*?

2 Name four secondary infections which most commonly complicate visceral leishmaniasis.

3 Anaemia is a common feature of visceral leishmaniasis. Is it:
   (a) Megaloblastic?
   (b) Normochromic and normocytic?
   (c) Due to marrow suppression or replacement?
   (d) Haemolytic?
   (e) Due to iron deficiency?

4 List three physical signs that may result from thrombocytopenia in kala-azar.

5 Describe the changes in plasma proteins produced by visceral leishmaniasis.

6 List the two main broad clinical ill-effects of these changes in plasma protein synthesis.

7 A 26-year-old English VSO (Voluntary Service Overseas) teacher has just returned from 18 months' teaching in northern Kenya. For 6 months he has suffered from bouts of irregular fever and sweating, recently accompanied by increasing weakness and breathlessness. On examination he is afebrile but pale, and his liver and spleen are both considerably enlarged. List the three most important investigations you would request.

8 A 3-year-old girl, daughter of an English mother and Greek father, returned 3 weeks ago from a 6-week holiday in a village near Athens. For 1 week she had been feverish and her spleen enlarged. Blood films are negative for parasites; her WBC is low ($2100/\mu l$) and platelet count so low ($15\,000/\mu l$) that her paediatrician does not think that spleen puncture would be safe. What would you advise? (She has already had repeated negative blood cultures, with the specific possibility of a *Salmonella* infection in mind.)

FURTHER READING

Control of the Leishmaniases (1990) Technical Report Series 793. World Health Organization, Geneva.

Molyneux, D.H., Ashford, R.W. (1983) *The Biology of Trypanosoma and Leishmania Parasites in Man and Domestic Animals*. Taylor & Francis, London.

Peters, W., Killick-Kendrick, R. (eds) (1987) *The Leishmaniases in Biology and Medicine*. Academic Press, London.

African trypanosomiasis

African trypanosomiasis is caused by species of *Trypanosoma brucei*. There are three morphologically identical parasite species:

1 *T. brucei brucei*, confined to domestic and wild animals.

2 *T. brucei gambiense*, causing gambiense sleeping sickness in west and central Africa.

3 *T. brucei rhodesiense*, causing rhodesiense sleeping sickness in east and southern Africa.

Despite their names, the disease caused by the organisms does not always cause sleepiness. Transmission is by the bite of tsetse flies (members of the genus *Glossina*) and the flies are only found in Africa. In general, the infected areas are found south of the Sahara and north of the Zambezi (Fig. 3.1).

PARASITES

The parasites forming the *T. brucei* group are morphologically indistinguishable. The parasites are flattened and fusiform in shape, like slender, pointed leaves 12–35 μm long and 1.5–3.5 μm broad. They are actively motile, using a thin, fin-like extension from the main body, the undulating membrane, to propel themselves. The form of the parasite found in humans is the trypomastigote, in which the kinetoplast is posterior to the nucleus, and from which the flagellum arises. The flagellum runs along the free edge of the undulating membrane, and usually projects in front of it, sometimes extending as far again as the creature's body. When some trypomastigotes have a free flagellum and some do not, as in the *T. brucei* group, the parasite is said to be dimorphic. The parasites multiply in the vertebrate host by simple fission, and dividing forms are often found in the blood. The parasites show great variation in their general configuration, some being short and sturdy (stumpy forms) and some long and slender (gracile).

LIFE CYCLE

This is the same for both species (Fig. 3.2), and infection is usually from

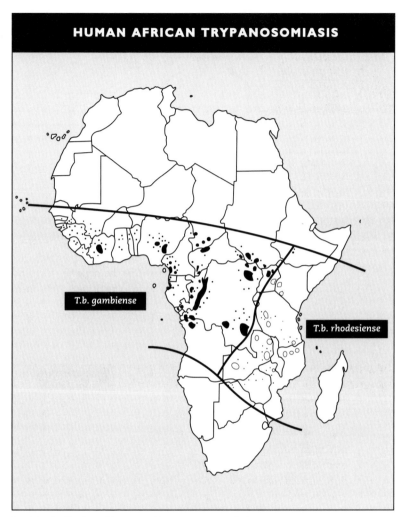

Fig. 3.1 Distribution of human African trypanosomiasis.

the bite of an infected tsetse fly in which the parasite has developed cyclically. It is likely that mechanical transmission via the proboscis of the tsetse sometimes occurs, and congenital infection has also been reported. Trypomastigotes from the infected host are taken up by the tsetse fly during a blood meal; both sexes of tsetse feed on blood only. In the stomach of the fly the parasites multiply by simple fission, penetrate the gut wall and migrate to the salivary glands by a rather complicated route. There the morphology changes, the kinetoplast coming to lie just in front of the nucleus, and the creatures are now called

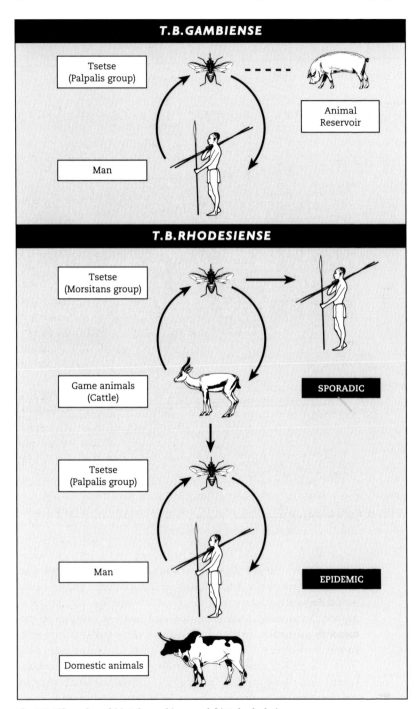

Fig. 3.2 Life cycles of (a) *T. b. gambiense* and (b) *T. b. rhodesiense*.

epimastigotes (crithidia). The infective trypomastigote (the metacyclic trypanosome) is found in the saliva about 20 days after the original infecting blood meal, and the fly remains infective throughout its normal life span of several months.

DISEASE

Initial response

Metacyclic trypanosomes injected during feeding initiate the infection. They commonly remain in the immediate area of the bite, actively dividing by simple fission for a variable period before becoming disseminated by the blood stream. Even the bite from an uninfected tsetse may cause considerable pain and a small, indurated lesion which persists for a few days. This is a bite reaction.

The local multiplication of trypanosomes may cause a much more marked inflammatory reaction—the trypanosomal chancre or bite lesion. This is a hot, raised, inflamed area, usually on the extremities and resembling a blind boil. It may cause enough pain to make the patient seek medical aid, much to his or her advantage if it is diagnosed correctly. On the other hand it may be relatively inconspicuous or rarely observed in *T. b. gambiense* infections.

The trypanosomal chancre appears 3 or more days after the bite and typically increases in size for 2 or 3 weeks, at the end of which time it begins to regress. Trypanosomes are found in the blood soon after the development of the chancre. In the more severe lesions, vesicles may appear over the chancre before it begins to regress. The local lymph glands may enlarge also.

Invasion: general features

The trypanosomes enter the blood from the chancre, via the lymphatics or perhaps directly, if the tsetse fed from a blood vessel rather than from a pool of blood caused by probing. When the trypanosomes enter the blood, the patient usually develops a fever. The pattern of fever varies with the parasite involved, but the host's reaction to the presence of the parasites is so variable that in the individual case the severity of the toxaemia cannot be used to distinguish between infection with one parasite or another.

In general, *T. b. gambiense* is better adapted to the human host than is *T. b. rhodesiense*. For this reason, *T. b. gambiense* is relatively well tolerated: the illness it causes tends to be subacute or chronic and parasitaemia may even be asymptomatic.

In contrast, *T. b. rhodesiense* is normally a zoonotic infection which is transmitted to humans 'accidentally'. It causes pronounced toxic effects: parasitaemia usually causes severe incapacity and the course of the illness is relatively rapid.

Immune response in the host

It might be supposed that the presence of active, replicating parasites in the blood and tissues would excite a vigorous immune response, and this is a prominent feature of both infections. The main response is antibody production. After initial infection, the parasitaemia typically rises, followed a little later by antibody production, mainly of IgM. When antibody has reached a level effective in destroying the parasites (the lytic process involves complement), parasitaemia rapidly falls and the parasites apparently disappear from the circulation. Later, parasitaemia recurs, but this generation of parasites has a different antigenic make-up from the previous population, and so is immune to lysis by the pre-existing antibodies. Typically, this second wave of parasitaemia is also terminated by antibody directed against it, and the whole process is repeated many times. The process by which selective pressure by antibody on the parasite population is followed by a change in antigenic make-up is called antigenic variation. It is a feature of both *T. b. gambiense* and *T. b. rhodesiense* infections, and probably this explains the fluctuating nature of the illness seen in many patients. The source of this reinfection of the blood has long been debated, but it now seems that cryptic trypanosomes are hidden away in the ependymal cells of the choroid plexus and reinfect the blood from there.

Effects of immune complex and complement

It seems likely that many features of the illness are due to complement activation. These would include capillary damage leading to localized areas of oedema, specific organ damage and the occasional occurrence of erythema nodosum, depending on the site of activation of the complement pathway. Higher levels of IgM are found in trypanosomiasis than in any other protozoal infection, and only some of this will react with trypanosomes. Later in the infection, IgG levels also rise, but not to the same extent as in malaria. This diversion of the body's humoral defences to a useless end inevitably impairs the response to other pathogens, mainly bacteria. Just as in visceral leishmaniasis, the infected host becomes much more susceptible to intercurrent infections.

CLINICAL PICTURE

T. b. gambiense: the early stage

Fever is the main early symptom, accompanied by headache and sometimes by fleeting areas of cutaneous oedema. The lymph glands are
invaded by the parasites, and become enlarged, the most prominent
glands often being in the posterior triangle of the neck. The presence
of enlarged glands in this site caused by *T. b. gambiense* is called
Winterbottom's sign.

In *T. b. gambiense*, the parasitaemia is not always of long duration, and
trypanosomes may be difficult to find in the blood, especially in late
infections. They are, however, found in the glands long after they have
apparently left the blood. The longer the duration of infection, the more
difficult it is to find trypanosomes anywhere.

The course of the fever is irregular, intermittent and chronic. The
spleen enlarges to a moderate size in many cases, and there may be a mild
normochromic anaemia. There are no obvious changes in the WBC,
but the erythrocyte sedimentation rate (ESR) is usually raised above
50 mm/h and sometimes to over 100 mm/h.

The early stage of the disease means the stage preceding involvement
of the CNS. Odd skin rashes sometimes occur (visible only in relatively
unpigmented skins), usually taking the form of areas of circinate erythema. There may be generalized pruritus, and characteristic thickening
of the facial tissues giving a sad or strangely expressionless appearance.
The duration of this early stage is usually many months, or even more
than 2 years. Occasionally patients are found in endemic areas with
asymptomatic parasitaemia. It is not ethical to leave them untreated,
however, as one cannot be sure they would not eventually succumb to
the infection.

Occasionally, patients with *T. b. gambiense* develop a rapidly progressive toxaemic disease which is fatal, as *T. b. rhodesiense* infection often is,
before the CNS is involved. Sometimes patients in the early stage die of
intercurrent infection, but usually death follows in the late stage, when
the CNS is invaded.

T. b. gambiense: the late stage

It is this stage which is responsible for the name sleeping sickness.
Trypanosomes in the brain give rise to chronic meningoencephalitis,
with cuffing of the small blood vessels by inflammatory cells (mainly
lymphocytes), the development of small granulomata in the brain, and a

degree of cerebral oedema. Cells with slightly refractile eosinophilic inclusions (Mott's morula cells) are often found in the brain substance. Neuronal destruction and patchy demyelination occur in advanced meningoencephalitis. Despite the severity of the symptoms in advanced late cases, the degree of functional recovery after successful chemo- therapy is remarkable.

Symptoms and signs in the late stage

The main symptoms are related to disturbed cerebral function. The first signs are perceived by the patient's relatives, and are those of an early organic dementia. The essential feature is a change in behaviour. A patient whose personal habits were previously fastidious becomes care- less about appearance; his or her speech becomes coarse and temper unpredictable and he or she may behave in a socially unacceptable way.

In one of DRB's patients, the first time his family noticed that something was amiss was when he left the table during a meal and urinated in the corner of the room, returning to his place as if nothing untoward had happened.

The psychiatric presentation may be so gross that the patient is sent to a psychiatric hospital, sometimes in a violent and excitable state resembling mania. Delusions may occur, and sometimes schizophrenia is simulated. Sleep becomes disordered, in that the patient sleeps badly at night but falls asleep during the day.

In the early evolution of this change, the patient can be readily awoken, and responds by conversing fairly normally. As time goes by, sleeping periods may become longer, until the patient is sleeping most of the time, and may even fall asleep while eating. At this stage, speech and motor functions in general are usually severely disturbed. When a patient is in this condition, the state of nutrition is governed by the efforts made to look after him or her by family or friends. If the patient is awoken and helped with feeding, body weight is usually fairly well maintained. If left alone, the patient will eventually starve to death.

If the 'sleeper' is well-cared-for by loving relatives, he or she eventu- ally slips into coma. Focal CNS signs may develop, but there is usually more diffuse evidence of CNS disease, especially relating to extra- pyramidal and cerebellar functions—widespread tremors involving the limbs, tongue and head, spasticity (mainly of the lower limbs), ataxia and sometimes choreiform movements. Convulsions are relatively uncom- mon. An odd physical sign that is sometimes seen before coma occurs is Kerandel's sign, in which, following firm pressure on the tissues overlying a bone such as the carpus or tibia, there is a definite delay before the patient shows any sign of pain.

In advanced cases, the tendon reflexes are often grossly exaggerated and the plantar responses may be extensor. In keeping with the inflammatory process in the meninges, the CSF shows an increase in cells, mainly lymphocytes, and protein. Death usually occurs within a few months of CNS involvement becoming manifest, but may be delayed for up to a year.

Clinical picture in *T. b. rhodesiense*

The parasite usually produces a more acute and virulent infection than does *T. b. gambiense*, and death from myocarditis may occur in the acute toxaemic stage before the CNS is involved. When the CNS is involved, the clinical features resemble the same stage in *T. b. gambiense*, but death is more rapid and neurological features are more pronounced than behavioural changes. *T. b. rhodesiense* may be fatal within a few weeks of the onset. In the early stages, *T. b. rhodesiense* may cause hepatocellular jaundice and mild anaemia, and severe anaemia may soon develop. Both liver and spleen may be slightly enlarged, and lymph gland enlargement (seldom so prominent as in *T. b. gambiense*) is commonest in the inguinal, axilliary and epitrochlear glands.

Serous effusions, especially into the pleural and pericardial spaces, are common and myocarditis occurs. Myocardial involvement is the commonest cause of death in the early stages. The picture in the late stage of *T. b. rhodesiense* infection is much like that of *T. b. gambiense* but occurs early in the course of the disease and is more rapidly progressive.

DIAGNOSIS

Direct methods

1 Blood films, stained with a Romanowsky stain as for malaria. Thick films produce a 10–20× concentration factor. Films can be examined wet (simply place a coverslip over a drop of blood) when the disturbance of the red cells produced by the movement of the trypanosomes can be detected using a dry ×40 objective.
2 Other methods using blood. Various concentration methods are used to detect scanty parasitaemia, best among which are microscopy of the buffy coat following centrifugation using the microhaematocrit (MHCT) and the recently developed, more sensitive, QBC technique as well as the minianion exchange column technique (MAEC).

Examination of blood films is most useful for *T. b. rhodesiense* infection. The organisms may also be isolated by inoculation into special culture media or into animals.

3 Gland puncture. This is of most use in *T. b. gambiense* where posterior cervical lymphadenopathy is common. A small-bore needle is inserted into an enlarged node held between thumb and finger. The flow of gland juice up the needle can be improved by massaging the gland while the needle is *in situ*. The juice is then expressed on to a slide and examined immediately. Some air in the syringe initially helps the procedure.

4 Bone marrow aspiration. This is useful in the early stages when other methods are negative.

5 CSF examination. The deposit from 5–10 ml of centrifuged fluid should be examined as soon as possible for motile trypanosomes or made into a smear, dried, fixed and stained with a Romanowsky stain.

6 The chancre. Trypanosomes can be recovered by aspiration from the chancre or from the regional glands draining the chancre, if they are enlarged, before the blood is positive.

Indirect diagnostic methods

Several serodiagnostic procedures have been used to detect antitrypanosomal antibodies, and such methods may be useful for screening suspects or when all attempts at direct diagnosis have failed.

The card agglutination test for trypanosomiasis (CATT) is technically simple to carry out, and gives good results in most areas of *T. b. gambiense* but is of no value in *T. b rhodesiense*. It is a valuable test for screening populations as the results are obtained within 30 min and many patients with *T. b. gambiense* are asymptomatic. The disadvantages at present are the problems of sensitivity and specificity of the antigen, as trypanosomes share antigens with several other protozoa and bacteria. Specific immunodiagnostic techniques are seldom available in countries where the disease is endemic.

Circumstantial diagnostic methods

The most useful is the detection of very high IgM levels, in both the blood and CSF, by immunodiffusion plates. Serum IgM in established trypanosomiasis is usually 8–16 times the normal maximum level, but many false-positive and false-negative results occur.

Is the infection *T. b. gambiense* or *T. b. rhodesiense*?

Biochemical taxonomic methods have been developed which help to differentiate the various zymodemes that constitute *T. b. gambiense* and *T. b. rhodesiense*. Two distinct groupings of enzyme types occur in *T. b. rhodesiense* (*busoga* and *zambezi* groups).

In the absence of such techniques, the most reliable method depends on the infection of albino rats: *T. b. rhodesiense* produces a fatal infection in 20–50 days; *T. b. gambiense* either fails to produce infection at all, or leads to a chronic, non-fatal infection which lasts for hundreds of days. This method is of no use to the individual patient, but may be of importance to the medical authorities in an area where there is an epidemic of sleeping sickness. The measures taken to control the epidemic will vary greatly according to the species involved.

Is the CNS involved?

If there are obvious physical signs, the question is already answered. Where there are no obvious signs, the question must be answered with certainty. The reason is that drugs effective only against the parasites outside the CNS will fail to cure a patient with CNS involvement. Initial improvement will then be followed by delayed relapse, in which CNS damage may be irreversible by the time the patient presents for further treatment.

The answer to the question depends on examination of the CSF. If blood examination is positive, lumbar puncture should be preceded by treatment to eliminate parasites in the blood, otherwise CSF infection could be introduced by the puncture itself. It is usual to wait 24 h after an injection of suramin before performing lumbar puncture. A rise in either cells or protein (or both) indicates CSF involvement. A cell count (in an atraumatic tap) of more than 5/μl indicates CNS involvement; the level set for significant protein levels depends on the method used, and should be related to the locally established norm. The upper limit of normal lies between 22 and 40 mg/dl. A rise of CSF IgM levels also indicates CNS disease.

TREATMENT

Early *T. b. gambiense* and *T. b. rhodesiense* infections

In almost all cases the trypomastigotes of both species can be eliminated from everywhere in the body, apart from the CNS, by suramin. Pentamidine is only effective in *T. b. gambiense*.

SURAMIN

This is the drug of choice in both species of trypanosome infection when the CSF is normal. It is a complex urea derivative with a high molecular weight, and is mainly bound to plasma proteins, with a correspondingly extended half-life. It must be given intravenously. It is available in 1 g ampoules of dry powder which must be dissolved in distilled water immediately before injection. A full adult course is 1 g intravenously on days 0, 3, 7, 14 and 21, following a test dose of 200 mg. Some patients develop a Herxheimer-like reaction after the first dose. This usually takes the form of fever, for which symptomatic treatment alone is sufficient.

When used to treat trypanosomiasis the drug is usually well tolerated. It does irritate the kidneys and almost always causes albuminuria. It is only necessary to defer a dose if heavy albuminuria occurs.

African trypanosomiasis with CNS involvement

The principles of treatment are:

1 Improve the general condition by treating anaemia, infection and malnutrition.

2 Give suramin to eliminate parasites outside the CNS.

3 Destroy the parasites in the brain with melarsoprol.

Melarsoprol (Mel B)

This is a trivalent arsenic compound in which melarsen oxide is combined with dimercaprol (BAL; its antidote) so as to minimize toxic effects, in solution with propylene glycol. It is given in courses designed to achieve the maximum therapeutic:toxic ratio. It is active against blood, tissue and CNS trypanosomes. Until melarsoprol was invented by Dr Ernst Friedheim, there was no effective treatment for *T. b. rhodesiense* infections with well-established CNS involvement. It is now the drug of choice for all cases of African trypanosomiasis where the parasite has invaded the CNS. The drug must be given intravenously.

DOSE

The safety margin is low. Five millilitre ampoules of a 3.6% solution in propylene glycol are standard, and the normal maximum dose is 3.6 mg/kg per day. The formulation has been devised to make dosage calculations easy: using this concentration, the maximum dose is 1 ml of solution per 10 kg body weight (0.1 ml/kg) to a maximum single dose of 5 ml. Initial courses start with low doses and increase in subsequent courses. The frequency of administration and total dosage are governed by various factors. Both late *T. b. gambiense* cases and late *T. b. rhodesiense* cases with CNS involvement are treated with melarsoprol—three or four courses, each of 3 days' duration separated by a rest period of 7 days without arsenical treatment, to a total of 35 ml melarsoprol. A commonly used treatment schedule in *T. b. rhodesiense* is shown in Table 3.1.

TREATMENT SCHEDULE			
Day	Drug	Volume (ml)	Volume (mg/kg)
1	Suramin	2.5	5
3		5.0	10
5		10.0	20
7	Melarsoprol	0.5	0.36
8		1.0	0.72
9		1.5	1.1
16	Melarsoprol	2.0	1.4
17		2.5	1.8
18		3.0	2.2
25	Melarsoprol	3.0	2.2
26		4.0	2.9
27		5.0	3.6
34	Melarsoprol	5.0	3.6
35		5.0	3.6
36		5.0	3.6

Table 3.1 Treatment schedule for an adult with late-stage T. b. rhodesiense.

The importance of prior treatment with suramin is well established in cases of T. b. rhodesiense infection. It is reasonable to wait 1 or 2 weeks after the suramin before embarking on a melarsoprol course in T. b. gambiense.

PRECAUTIONS AND TOXIC EFFECTS

Melarsoprol, like all trivalent arsenical drugs, may cause a serious or even fatal encephalopathy (reactive arsenical encephalopathy). Its use is only justified in a disease as grave as cerebral trypanosomiasis, where the outlook for the untreated disease is hopeless. It has to be accepted that 1–5% of patients treated with it will die of the effects of the drug itself.

The danger of severe toxic effects is minimized by improving the patient's general condition, as already described.

If arsenical encephalopathy does occur, injections of dimercaprol (BAL) may help. It should be remembered that dimercaprol can itself be toxic.

Difluoromethyl ornithine (DFMO) at a dose of 400 mg/kg daily, initially intravenously and then orally, for up to 6 weeks is effective in T. b. gambiense, including arsenic-resistant cases, but appears to be ineffective in T. b. rhodesiense. It is expensive and difficult to administer. Side-effects (gastrointestinal symptoms and anaemia) do not usually require treatment to be stopped.

Monitoring cure

Not only should the patient's symptoms resolve after treatment, but the CSF should slowly return to normal. The IgM levels in the blood should also return to normal within 6 months of treatment.

It may take 6 months or more for the CSF cell count to fall below $5/\mu l$ and the protein below $25-40\,mg/dl$, depending on the method used. Failure of these parameters to reach normal may be the first indication that treatment has been unsuccessful. Full cure cannot be assumed unless a 2-year follow-up has been completed. If treatment of patients with CNS involvement has been delayed, a variable degree of neurological defect will persist. Most commonly this takes the form of intellectual impairment.

EPIDEMIOLOGY

T. b. gambiense infection

Although the pig and other animals have been found to be naturally infected in some parts of west Africa, humans are much the most important reservoir of infection. Infection is spread from human to human by the bite of riverine tsetse flies (*palpalis* group) which breed along the banks of rivers and lakes: they require protection from direct sunlight and need a high humidity to thrive. Infection tends to occur where human activities bring humans into contact with the fly, such as at river crossings and sites used for collection of water, and when fishermen come into contact with flies on the river or lake shores. Village-sized and larger outbreaks occur, sometimes amounting to epidemics. The spread of epidemics tends to be linear, following the distribution of flies along the course of rivers, or affecting islands in lakes.

Principles of control of *T. b. gambiense*

1 Reduction of human–fly contact. This can be achieved by felling trees around water contact points 20 m each side of the bank and for 200 m up and downstream of the crossing. If trees are simply felled and left to lie across the stream or river, they will soon lose their leaves and with them their capacity to provide shade. The local population will soon then use the remaining wood for fuel.

2 Active case-finding and treatment. This is very important in *T. b. gambiense* outbreaks, and its effectiveness has been thoroughly proved by the trypanosomiasis control units in west Africa.

3 A direct attack on the adult flies using insecticides. This can achieve a very rapid reduction in tsetse populations, but is expensive, and requires a high degree of supervision and a very disciplined programme if it is to be effective. The most rapidly effective high-technology method is aerial spraying of insecticide.

4 The widespread use of tsetse traps, such as the pyramidal trap which can be impregnated with insecticide and placed at points of high human–fly contact.

T. b. rhodesiense infection

T. b. rhodesiense is usually a zoonotic infection in members of the antelope family, most especially the bushbuck. It can be maintained as a zoonosis in the animal population in the absence of human cases, and is transmitted by tsetse species dwelling in savannah and woodland habitats (morsitans group). It is an especial hazard to those who spend long periods in enzootic areas in pursuit of their livelihood there, such as hunters and honey-gatherers. A more recent group at risk is the growing band of tourists intent on seeing African game in its natural habitat.

Although T. b. rhodesiense cases tend to be sporadic, epidemics do occur, especially in east Africa around Lake Victoria. In these epidemics tsetse populations build up adjacent to human populations containing active cases. The usual morsitans group vector is replaced with a palpalis group tsetse. Domestic cattle are infected, develop a chronic parasitaemia and act as a reservoir host. The current epidemic in south-east Uganda is caused by peridomestic breeding Glossina fuscipes in thickets of the exotic plant Lantana camora.

Principles of control of T. b. rhodesiense

1 Reduction of human–fly contact. This is difficult in view of its sporadic nature. There is no doubt that intensive agricultural development of an area, with its inevitable tree-felling and destruction of natural tsetse breeding habitats, helps to eliminate the disease. Those most at risk of exposure are farmers living at the edge of a development scheme engaged in the pioneering work of marginal agriculture.

2 Active case-finding and treatment will only make much of a contribution to control when human-to-human transmission is occurring in an epidemic but may also help in epidemics to detect cases early. Where the reservoir of infection is in the wild animal population, selective game destruction has been used. Domestic animal population can be treated with trypanocidal drugs.

3 A direct attack on the adult flies using residual insecticide spraying or aerial spraying is only feasible where a large economic reward is in sight, such as by making a tsetse area completely free of the fly, and opening up the country to the possibility of cattle ranching. This has been successfully done in parts of southern Africa. When it is done, reinvasion of the area by tsetses from adjacent untreated areas must be prevented by the creation of broad areas completely free of vegetation–tsetse barriers. The tendency of flies to rest on the undersurfaces of vehicles and be passively transmitted long distances and so possibly recolonize a cleared area can be anticipated by setting up barrier posts at which all vehicles (including road vehicles and railway trains) are stopped and sprayed with insecticide.

4 Breeding sites of forest and savannah tsetse can be eliminated by felling the shade trees beneath which they breed. This is only economically feasible where only some of the trees are needed by the flies. Indiscriminate felling may convert fertile forest into desert. Some ecologists think that, by preventing economic cattle ranching in infected areas, the tsetse has played an important role in preventing desertification. For this reason the tsetse has been called the shield of Africa.

5 The use of tsetse traps and deltamethrin-impregnated screens or targets has been found to be highly effective in several parts of Africa. In favourable conditions flies may be completely 'trapped out'.

AFRICAN TRYPANOSOMIASIS: SURVEILLANCE AND NOTIFICATION

In all endemic areas the disease should be made notifiable to a central trypanosomiasis control unit. Specialized staff can then be sent promptly to the area and steps taken (such as active case-finding and treatment) to prevent the development of an epidemic. This strategy has proved very effective in Ghana, Nigeria and Uganda.

The breakdown of such a system due to civil disturbances may rapidly lead to the development of epidemics. Zaïre has provided a good example of this in recent years.

QUESTIONS, PROBLEMS AND CASES

1 Give a general account of the differences in the clinical pictures of *T. b. gambiense* and *T. b. rhodesiense* in humans.

2 What are the morphological differences between *T. b. gambiense* and *T. b. rhodesiense*?

3 What is the best direct diagnostic test for *T. b. gambiense* infection?

4 What is the best single diagnostic test for *T. b. rhodesiense* infection?

5 You suspect *T. b. gambiense* infection in a west African with negative blood films and gland puncture smears. What other sites may reveal trypanosomes?

6 What changes in the serum proteins are highly suggestive of an active African trypanosome infection?

7 A febrile adult patient from west Africa is stuporose, and has an abnormal CSF with raised cells (50/μl) and protein levels (60 mg/dl). Most of the cells are lymphocytes. The following diagnoses are worth considering seriously — mark each item T (true) or F (false):
  (a) Meningococcal meningitis.
  (b) *T. b. rhodesiense* sleeping sickness.
  (c) Meningovascular syphilis.
  (d) Cerebral malaria.
  (e) *T. b. gambiense* sleeping sickness.
  (f) Cryptococcal meningitis.
  (g) A viral meningoencephalitis.

8 The following skin changes may be seen in African trypanosomiasis — mark T or F:
  (a) Diffuse pruritus.
  (b) A nodular or vesicular eruption.
  (c) Fleeting patches of oedema.
  (d) Transient circinate erythematous lesions.

9 *T. b. rhodesiense* and *T. b. gambiense* infections can be distinguished from each other by — mark T or F:
  (a) The clinical picture in the individual patient.
  (b) The course of the illness in inoculated albino rats.
  (c) Morphological differences developing in culture.
  (d) Reaction with specific antibodies.

10 Melarsoprol is active against all stages of African trypanosomiasis. Why is it not used for treating all stages of the disease?

11 List three ways in which African trypanosomiasis can be transmitted.

12 What precautions should be taken before performing lumbar puncture in a case of suspected CNS involvement in a patient with African trypanosomiasis?

13 There has been a localized outbreak of *T. b. gambiense* sleeping sickness in a village near your hospital. Assuming you had the Administrative Medical Officer's permission, what could you do to stop the outbreak?

14 The following syndromes/associations would make you suspect African sleeping sickness – mark T or F:

(a) A febrile patient with facial oedema and itching.

(b) A patient with a history of several weeks' fever and hepatocellular jaundice.

(c) A violently excited patient with fever and lymphadenopathy.

(d) A febrile patient who seems in reasonably good condition apart from complaining of headache, but who has a sullen expression.

FURTHER READING

Epidemiology and Control of African Trypanosomiasis. (1986) Technical Report Series 793. World Health Organization, Geneva.

Mulligan, H.W. (ed.) (1970) *The African Trypanosomiases*. George Allen and Unwin, London. [This is an old publication but contains many interesting chapters, especially regarding the historical aspects of trypanosomiasis.]

South American trypanosomiasis (Chagas' disease)

South American trypanosomiasis is found in humans and animals and is widespread in central and south America. It is caused by *Trypanosoma cruzi*. Infection occurs in animals as far north as Virginia, and in humans in Texas. Trypanosomes appear in the blood of the mammalian host. They differ from trypomastigotes of the *T. brucei* group in having a larger kinetoplast. They are taken up by reduviid bugs (assassin bugs, kissing bugs) which bite at night. All stages feed on blood, but only adult bugs can fly. Organisms multiply in the hindgut of the bug as epimastigotes. Infection is acquired by rubbing into the bite wound or conjunctiva metacyclic trypanosomes in the faeces voided by the bug during feeding. Infection can also be acquired by blood transfusion and congenital infections are common.

In the host, trypomastigotes enter a variety of tissue cells and multiply exclusively as intracellular amastigotes. Liberated from ruptured host cells, amastigotes appear as trypomastigotes in the blood and reinvade new cells.

CLINICAL FEATURES

There may be initial orbital oedema (Romaña's sign) or an area of cutaneous oedema (a chagoma) elsewhere, followed in a small proportion by a toxaemic stage with parasites in the blood, in which lymph glands, liver and spleen may be enlarged. Death sometimes occurs in this stage due to cardiac damage or meningoencephalitis, especially in children. Then follows a long latent period in which intermittent low parasitaemia occurs, but intracellular replication as amastigotes continues.

Asymptomatic infection is common.

Years after initial infection, complications may develop, mainly:

I Cardiomegaly due to destruction of myocardial fibres and ganglion cells. This presents mainly as:

(a) A congestive cardiomyopathy with cardiac failure.

(b) Cardiac arrhythmias. Sudden death may occur.

2 Megaoesophagus due to destruction of the intramural plexus. This commonly presents as aspiration pneumonia.

3 Megacolon, also due to destruction of myenteric plexus. This presents as intractable constipation and abdominal distension.

4 Similar 'mega' disorders of other hollow muscular viscera, such as the small bowel and ureters.

DIAGNOSIS

Direct diagnosis

1 By finding trypanosomes in blood films (various concentration techniques are employed). This is of value in early acute infection.

2 By xenodiagnosis to detect very low parasitaemia: 'clean' bugs raised in the laboratory are fed on the patient. Some 2–4 weeks later they are killed, and the gut examined for the presence of epimastigotes (crithidia) or metacyclic trypanosomes.

3 By demonstrating amastigotes in a muscle biopsy specimen (seldom used).

4 By blood culture on NNN medium, when epimastigotes appear in positive cases.

Indirect diagnosis

There are many effective methods which allow the detection of antibody in active cases without parasitaemia, such as the complement fixation test and IFAT.

TREATMENT

Several drugs can terminate the parasitaemia in the acute toxaemic stage, such as nifurtimox (Lampit), various other nitrofurazones and primaquine. No drug has proved effective in eradicating the intracellular amastigotes.

EPIDEMIOLOGY AND CONTROL

Human infections only become important when peridomestic bugs become established in people's houses. A single adobe hut may harbour more than 4000 bugs. Although the bugs are susceptible to insecticides, long-term control requires improving the standard of housing in endemic areas. The most important changes would be the elimination of cracks in mud walls by rendering with cement, and the replacement of thatch or palm-leaf roofing with metal sheets. In many parts of the Americas where the disease is enzootic, human infections do not occur because the reduviid vectors will not bite humans.

Relapsing fever

Relapsing fever in a febrile illness caused by spiral organisms of the genus *Borrelia*, 6–10 μm long with 5–10 'turns'. The organism differs from spirochaetes in staining readily with Romanowsky stains. It multiplies by simple fission. Antigenic variation is responsible for the typical relapses.

LOUSE-BORNE RELAPSING FEVER (*Borrelia recurrentis*)

Transmission

This is by the body louse *Pediculus humanus* (rarely by head lice), when infected louse body fluids enter the bite wound or abrasions after it is crushed by scratching. It is found in almost any part of the world where people live in squalor but are clothed.

Reservoir of infection

Human beings.

TICK-BORNE RELAPSING FEVER

This is caused by *B. duttoni* in Africa, and other species elsewhere.

Transmission

This is by the bite or body fluids of soft ticks of the genus *Ornithodoros*.

Reservoir of infection

The reservoir of infection is in rodents and the ticks themselves. Ticks are infected for life and can transmit infection for five generations through their eggs. The disease is present in Africa, southern Europe, the Middle East, Asia and the Americas.

DISEASE

The two types of infection differ in detail only. Common features are:

1 Rapid onset with fever, headache, prostration and myalgia, in the presence of bacteraemia, lasting 4–7 days.

2 The tendency to spontaneous remissions lasting a similar period, followed by:

3 Relapses, commonly one or two in louse-borne, three to six in tick-borne, and up to 11 in African infections.

4 The tendency to hypotensive collapse when the temperature falls rapidly.

ORGAN INVOLVEMENT

Myocarditis and meningoencephalitis are fairly common. Liver, spleen and lymph glands are often enlarged. Cough and sputum, with or without pneumonic signs, are also common. Hepatocellular jaundice, liver failure and DIC may complicate louse-borne disease. There is often a petechial rash in severe cases.

ENDEMIC TICK-BORNE RELAPSING FEVER

In east Africa infections in adults are often asymptomatic. Young children suffer a febrile illness and pregnant women have severe illnesses which may result in abortion, stillbirth or maternal death. Congenital transmission of infection occurs.

DIAGNOSIS

Specific diagnosis

This is by finding organisms in Romanowsky-stained thick blood films taken during a febrile episode. Organisms are more numerous in louse-borne cases. After inoculation of blood into laboratory mice, parasitaemia develops in 2–3 days. Various tests to demonstrate antibodies are available; false-positive reaginic tests for syphilis are common.

Blood changes

These are typically a high polymorphonuclear leucocytosis and mild anaemia. The platelet count is not usually low except when DIC occurs.

TREATMENT

A tetracycline in normal doses (1–2 g/day for 5–7 days) effectively terminates parasitaemia, but a Herxheimer-like reaction often follows, especially in louse-borne relapsing fever, with a rise in temperature, confusion, tachycardia and transient hypertension followed by hypotension. Intravenous meptazinol in a dose of 300–500 mg dramatically reduces the severity of the reaction. Supportive treatment with intravenous fluids and oxygen may be needed. Penicillin is usually effective, but resistance has been recorded.

EPIDEMIOLOGY AND CONTROL

Louse-borne relapsing fever may be endemic or epidemic, and epidemics are encouraged by war, famine and mass population movements. Epidemics affect people who wear clothes which provide refuge for the lice. This is a disease of squalor and human disaster.

The most rapid control measure is by mass delousing, using louse powder. Widespread resistance to organochlorine insecticides has led to their replacement by organophosphorous compounds such as malathion: 1% malathion dusting powder is effective and non-toxic.

Tick-borne relapsing fever affects a locality for years, due to an animal reservoir and/or the longevity and transovarial passage of infection in the tick.

Personal protection involves the use of a sleeping net, sleeping as high off the floor as possible, protective clothing and repellents. Infected localities can sometimes be freed of infection by a determined attempt to eradicate the ticks, such as by residual spraying with insecticide, perhaps combined with trapping of rodents.

FURTHER READING

Barclay, A.J.G., Coulter, J.B.S. (1990) Tick-borne relapsing fever in central Tanzania. Trans R Soc Trop Med Hyg 84, 852–6.
Goubau, P.F. (1984) Relapsing fevers; a review. Ann Soc Belge Med Trop 64, 335–64.

CHAPTER 6

Rickettsial infections

There are eight species of *Rickettsia* infecting humans, and a considerable number of local variants of the type species. They are of worldwide distribution. These are the features they have in common:

1 They are minute, pleomorphic, Gram-negative intracellular organisms.
2 They grow on chick yolk-sac and tissue culture.
3 They will infect rodents.
4 Their pathogenic effects come from replication in endothelial cells of small blood vessels and perivascular inflammation.
5 They are transmitted by the bites, body fluids or faeces of arthropods.
6 The illness is often characterized by fever and rash.
7 There is great variation in the severity of illness produced by each organism.

We shall deal in more detail only with the two rickettsial infections that are particularly tropical in the geographical sense: scrub typhus and African tick typhus.

SCRUB TYPHUS

This is also called mite typhus and tsutsugamushi fever. The organism is *R. orientalis*, a zoonosis of rodents transmitted by larval trombiculid mites. Transovarial passage of the infection can continue for several generations The reservoir of infection is in the mite–rodent complex. Humans are infected by the bite of an infected larval mite. The infection is widely distributed in South-east Asia, Oceania, northern parts of Australia, and extends as far west as South-east Siberia.

CLINICAL PICTURE

A papule forms at the site of infection which later usually becomes necrotic to form the typical black eschar. Four days to 2 weeks after the bite, the illness begins with fever and malaise, followed by adenitis in the glands draining the bite. As the organisms spread throughout the body, fever, malaise and headache increase and a general lymphadenopathy occurs in most cases. The spleen also enlarges. About a week after the onset the main features are:

1 Continuous fever.

2 Cough and signs of bronchitis or pneumonia.

3 Photophobia and conjunctivitis.

4 A maculopapular rash maximal over the trunk and proximal parts of the limbs, spreading to the extremities.

5 Generalized adenopathy.

6 Tender, localized adenopathy related to the bite.

7 Splenomegaly.

8 Delirium and deafness.

In severe cases the fever continues for 2 weeks before it begins to subside by lysis. In the most severe cases death occurs towards the end of the first week or during the second week. The patient dies in a toxaemic state, and the precise anatomical cause of death cannot usually be ascertained, although myocarditis and DIC both occur.

SPECTRUM OF SEVERITY

This is the typical sort of case one sees in a hospital in South-east Asia, presenting late and without prior treatment. Many cases of relatively minor illness occur, and many of them never seek medical aid. On the other hand, fulminating infections may cause death before the typical features have developed.

DIAGNOSIS

In an endemic area, the clinical picture is sufficiently distinctive for a clinical diagnosis to be made with a high degree of confidence. The most important supporting investigation is a neutrophil leucopenia.

Specific diagnosis is seldom possible early enough to help in management. Cell concentrates of blood taken during the early stages may cause a patent infection about 2 weeks after intraperitoneal injection into mice, but *Rickettsiae* are too dangerous to handle in routine laboratories. The Weil–Felix test usually becomes positive to OXK *Proteus* strains in the second week. About the same time, specific antibodies to *R. orientalis* may be demonstrated by immunofluorescence or immunoperoxidase tests, but only in specialist laboratories.

TREATMENT

Tetracycline is the drug of choice. The usual adult course starts with 2 g, and continues with 500 mg 6-hourly for 10 days. Chloramphenicol in similar doses is also effective.

EPIDEMIOLOGY AND CONTROL

The epidemiology varies in different parts of the world, but transmission sites are often associated with natural habitats modified by humans, as when the primeval jungle is felled and replaced by a secondary growth of scrub. Notorious areas for human infection are called mite islands, and can be destroyed by cutting down all vegetation and burning it. Because the organism is transmitted transovarially, the mites themselves act as reservoirs of infection, so no immediate effect is achieved by rodent control.

When backyard outbreaks occur in a town, the most rapid way to stop transmission is to spray the area with residual insecticide.

There is no vaccine. Soldiers and others who must enter infected areas can be protected to a great extent by using clothing impregnated with a repellent such as dimethyl phthalate, providing the repellent is renewed frequently. Doxycycline 200 mg once weekly has been used sucessfully for short-term prophylaxis.

AFRICAN TICK TYPHUS

The organism R. conorii var. pijperi is widespread in sub-Saharan Africa, and transmitted by various species of ixodid tick. The reservoir is probably in rodents. The infection is common in those exposed to tick bites on camping safari, and may also be brought into the home by infected dog ticks.

CLINICAL PICTURE

The disease resembles a very mild attack of scrub typhus, and an eschar and secondary adenitis are usual. A maculopapular rash starts on the trunk and spreads to the limbs; it sometimes becomes vesicular or petechial. Fever usually only lasts a few days, and the mortality is negligible.

Complications are rare, but scrotal gangrene has been reported, and the old and those with G6PD deficiency may suffer a severe illness.

DIAGNOSIS

The Weil–Felix test often shows antibodies to both OXK and OX19 strains of *Proteus*. Specific antirickettsial antibodies may be demonstrated in specialist laboratories. A biopsy of the eschar can be used to demonstrate rickettsiae by immunofluoresence.

TREATMENT

Some disease is so mild that treatment is scarcely worthwhile. In more severe cases, either tetracycline or chloramphenicol may be used.

PREVENTION

People camping in endemic areas should try to protect themselves from tick bite by using camp beds, protective clothing, repellents and inspecting skin surfaces for ticks which are removed at once. Domestic infection can be prevented by regular deticking of dogs. A dichlorvos-impregnated tick collar worn by the dog will help keep down the number of ticks brought into the house, provided it is renewed regularly.

RELATED DISEASES

Diseases with a similar clinical picture have been reported from many other parts of the world. In southern Europe the dog is the reservoir and the dog tick the vector, with transovarial passage of the infection.

American tick typhus caused by *R. rickettsi* occurs in Colombia and Brazil, and may cause severe infection with haemorrhagic or gangrenous rash, like that seen in Rocky Mountain spotted fever.

Q fever, caused by *Coxiella burnetii*, probably occurs worldwide. Although the natural reservoir is in wild animals, humans usually acquire the infection from close contact with infected stock animals rather than from ticks.

A full textbook should be consulted for details of these infections, and descriptions of epidemic (louse-borne) and endemic (flea-borne) typhus.

FURTHER READING

Brown, G.W., Shirai, A., Jegathesan, M. *et al.* (1984) Febrile illness in Malaysia—an analysis of 1269 hospitalised patients. *Am J Trop Med Hyg* **33**, 311–15.

Perine, P.L., Chandler, B.P., Krause, D.K. *et al.* (1992) A clinico-epidemiological study of epidemic typhus in Africa. *Clin Infect Dis* **14**, 1149–58.

Arbovirus and rodent-borne infections, and rabies

It is not possible to deal with tropical virus infections clearly, accurately and in a small space. This is partly because there are so many and because it is difficult to devise a useful classification. The virologist's classification into various arbovirus groups is of little help to the clinician. Classification by syndrome may also be confusing because the same virus may cause more than one syndrome. This is the compromise system I shall follow:

1 Arboviruses (arthropod-borne viruses).
2 Directly transmissible haemorrhagic viruses.
3 Rabies.

ARBOVIRUS INFECTIONS

Almost all these infections are zoonoses, in which the infection is normally maintained in a wild animal reservoir, commonly rodents and birds. In some of them, viral replication in a host living closer to humans (an amplifier host) has an important effect on human infection. A few are maintained in human hosts without the need for a persisting animal reservoir. The commonest vectors are mosquitoes, ticks and sandflies. Epidemics caused by biting flies tend to be explosive, and rapidly move through a community until the level of immunity is so high that transmission stops. The epidemic does not then recur until a sufficiently large susceptible population has built up again. But endemic transmission among the non-immunes may continue, and prevent the development of epidemics. In such an area one may be surrounded by infection, but never realize that it exists.

Acquired immunity to arbovirus infections is prolonged and often lifelong. Epidemics are associated with ecological change, especially land development, where vectors and vector abundance are altered or by the introduction of non-immune settlers. The arthropod vectors, once infective, are infective for life and transovarial transmission may occur. When the vector is a tick, the infection tends to persist in one place for a long time, because the ticks are relatively static and long-lived. When rodents are the reservoir, their territorial instincts confine them to a limited area, and this restricts the spread of infection also.

Arbovirus infections cause four clinical syndromes:

1 Acute undifferentiated fever.

2 Fever with a morbilliform rash.

3 Fever with hepatitis and haemorrhagic features.

4 Fever with encephalitis.

Many virus infections may present with several syndromes.

All arbovirus infections tend to run a biphasic course. In the early stages virus replication occurs within susceptible cells. In the second phase, which may follow a period of relative well-being, virus/antibody aggregates are deposited in blood vessels, and the more serious clinical effects often then occur as a consequence of vascular damage.

Dengue fever: a typical arbovirus fever

This is perhaps the most important arbovirus in which no significant animal reservoir has been found. It probably causes as much recognizable disease as all the other arboviruses put together. Large epidemics are due to transmission by the mosquito *Aedes aegypti* in densely populated, urban areas.

After an incubation period of 5–8 days there is a sudden onset of fever, headache, severe musculoskeletal pains and often upper respiratory symptoms. The fever sometimes falls temporarily after a few days, and then recurs. The typical rash usually develops on the third to fifth day, during the second half of the illness. It is usually maculopapular, starts on the trunk and then spreads to the limbs and face. A few days after the rash appears, the temperature begins to fall and recovery begins. There are no permanent sequelae, and provided haemorrhagic complications or shock do not occur, the mortality is very low. Diagnosis is by virus isolation or the demonstration of rising antibody titres in a virology laboratory. Treatment is supportive and symptomatic. There is no commercially available vaccine. Outbreaks are controlled by *Aedes* control, as for yellow fever.

The other fevers, listed in Table 7.1, can cause a similar or indistinguishable clinical picture, but a significant lymphadenopathy is more common in O'nyongnyong, Chikungunya and West Nile virus infections.

Yellow fever

Yellow fever is a zoonosis of monkeys in Africa and south and central America, transmitted from monkey to monkey by mosquitoes biting in the forest canopy. Humans can become infected by a bite from a mosquito in the forest—so-called sylvatic yellow fever. Epidemic, human-to-human transmission occurs when a viraemic patient enters a populated area, especially where the susceptible peridomestic mosquito, A. aegypti, abounds. These are the conditions needed for urban or epidemic yellow fever to develop. Epidemics also occur at the margins of yellow fever endemicity. Endemic and epidemic areas for yellow fever are shown in Fig. 7.1. Many infections are inapparent, and many only present as a mild, self-limiting fever. This is particularly common where other antigenically related (group B) arbovirus infections occur in the area. In classical cases (this does not mean typical, but florid), the following features may be present:

1 Initial fever and prostration with headache, generalized myalgia and abdominal pain and vomiting.

2 Sometimes a short remission then occurs, to be followed by vomiting, bleeding, shock, and signs of progressive hepatic and renal failure.

3 If the patient survives, complete recovery occurs.

Specific diagnosis requires virus identification in a specialist laboratory, or the demonstration of a rise in antibody titre in patients who recover. A postmortem liver specimen removed by viscerotome usually shows the characteristic mid-zone necrosis and Councilman bodies, but mistakes can be made as the changes are not pathognomonic.

Treatment is supportive and symptomatic.

The disease can be prevented by vaccination with living attenuated yellow fever virus of the 17D strain, which gives at least 10 years' protection. The International Certificate is valid for 10 years, beginning 10 days after vaccination.

An urban outbreak can be rapidly terminated by insecticide fogging to kill the adult Aedes, eliminating peridomestic breeding sites as far as possible, and adding temephos (Abate) to the water jars in people's houses which so often act as breeding sites. Epidemics can be prevented if the health authorities ensure the A. aegypti population is always kept under control or, failing that, making sure that immigrants from infected areas are vaccinated before entry.

Arboviruses causing fever, with or without rash

Table 7.1 (much abbreviated) gives an indication of the diversity of the problem.

Arboviruses and the haemorrhagic fever syndrome

The most important is dengue, although Chikungunya viruses, Rift Valley fever and the tick-borne Kyasanur forest fever in India may also cause

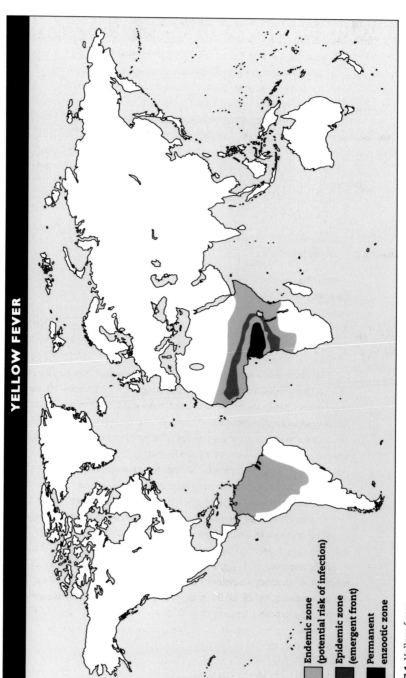

YELLOW FEVER

Endemic zone
(potential risk of infection)

Epidemic zone
(emergent front)

Permanent
enzootic zone

Fig. 7.1 Yellow fever.

DIVERSITY OF ARBOVIRUSES

Name of virus	Vector	Distribution
Dengue (four serotypes)	Mosquito (A. aegypti in epidemics)	Almost all tropical and many subtropical areas
Chikungunya	Mosquito	Tropical Africa and Asia
O'nyongnyong	Mosquito	sub-Saharan Africa
West Nile	Mosquito	Africa, extending to Egypt, South-east Asia, south France
Kyasanur forest	Ixodid tick	India
Phlebotomus fever	Sandflies	South Europe, north Africa, South-east Asia

Table 7.1 Diversity of arboviruses causing fever.

haemorrhage, and yellow fever also has haemorrhagic features. I shall use the commonest to illustrate the important features.

Dengue haemorrhagic fever (DHF) and dengue shock syndrome (DSS)

The disease has been reported sporadically for many decades, but the first recorded epidemic was in 1953 in Bangkok. Since then there have been many epidemics or smaller outbreaks reported from towns and rural areas in South-east Asia. Whenever there is an outbreak of DHF, the haemorrhagic cases are far outnumbered by cases with the dengue fever syndrome, or a nondescript mild fever.

The DHF syndrome is an unusual response to infection with the virus. The pathogenesis is not entirely understood, but there are clues: most of the cases occur in children (maximum incidence age 3–6 years), and in a large number of those infected there is serological evidence of previous exposure to dengue or another group B arbovirus infection. It has been proposed that reinfection with a different dengue strain accelerates and aggravates the second phase of the infection when virus–antibody complexes cause vascular damage. It would explain the extreme rarity of DHF in the fairly large number of expatriate children contracting uncomplicated dengue in areas where DHF is common in the indigenous population. A. aegypti has always been the important vector.

DHF: THE CLINICAL PICTURE

After the fever of 2–4 days' duration, accompanied by non-specific symptoms such as headache, abdominal pain, anorexia and vomiting, the child suddenly collapses with hypotension, a maculopapular rash and bleeding into the skin. Bleeding is seen as petechiae and bruising, epistaxis, haematemesis, melaena and haematuria. Shock is the worst prognostic sign, and requires the most urgent action. Other acute complications are pneumonia, myocarditis and pleural effusion. The major causes of bleeding are DIC and capillary damage. Postmortem examination reveals widespread vascular damage, haemorrhage and effusions, but except when gross cerebral haemorrhage is found there are no organ-specific changes sufficient to account for death.

There is no useful method for laboratory diagnosis except in retrospect, so the diagnosis must be clinical. The virus can only be isolated in the first 3 days of the illness. Clinical diagnosis is easy in the course of an epidemic, but very difficult between epidemics. But as treatment is supportive, the failure to make a correct diagnosis is mainly of public health importance. Treatment is directed towards restoring blood volume, by blood, plasma or plasma substitutes, and saline alone is better than nothing. The haemorrhagic state is usually rapidly self-limiting, but fresh, platelet-rich blood may help. There is some evidence that the careful use of heparin may reduce mortality, but this is inconclusive, and its dangers are well known. The value of corticosteroids is also not firmly established.

DHF can be prevented by effective control of A. aegypti. It is mainly a matter of effective application of well-established methods. There is no vaccine at present.

Arbovirus encephalitis

There are some arboviruses so well known for their tendency to cause encephalitis that their names reflect it. But in all these infections, for each patient who develops encephalitis, there are many with either inapparent infection or a non-specific febrile illness without encephalitis. So to have Japanese encephalitis virus infection does not mean you will necessarily suffer from encephalitis.

Some of the viruses causing encephalitis are listed in Table 7.2.

Japanese encephalitis

Japanese encephalitis occurs in epidemic or sporadic form over large areas of Asia and the western Pacific. The epidemiology is complicated but interesting. The main vectors are Culex mosquitoes which breed in rice fields. There may be several maintenance hosts such as birds (intense transmission occurs among young herons in Japan), bats or pigs. The

VIRUSES CAUSING ENCEPHALITIS

Name of virus	Vector	Distribution
Japanese encephalitis	Mosquito	Widespread in Asia and the Western Pacific
St Louis encephalitis	Mosquito	North America, Panama, Brazil, Argentina, the Caribbean
Venezuelan equine encephalitis	Mosquito	North and south America
Western equine encephalitis	Mosquito	North and south America

Table 7.2 Viruses causing encephalitis in tropical countries.

virus multiplication in the domestic pig brings a rich source of virus to our back door, and greatly increases the risk of human infection. The pig is then acting as an amplifier host.

In epidemics, children are the main victims. In Japan the ratio of encephalitis cases to infection may be as low as 1:1000 in children, but in American servicemen in Korea the ratio was about 1:25. A human epidemic is presaged by high infection rates found in pigs examined at slaughter.

There is usually a sudden onset with fever, headache and vomiting. Fits are common in children, as are disturbances of consciousness in all ages. A variety of pyramidal and extrapyramidal signs may develop. There may be meningism in the early stages, accompanied by pleocytosis and raised protein in the CSF. The CNS signs usually follow soon after the onset of fever. If the outcome is fatal, it usually occurs in the first 10 days.

Recovery may take months, and varying degrees of intellectual and neurological damage, often severe, may persist indefinitely. Specific virological diagnosis can only be carried out in a specialist laboratory.

Treatment is supportive only. In epidemics, control measures should be directed towards controlling the vector (usually *Culex tritaeniorhynchus*) and if acceptable to the population, the slaughter of pigs. An effective vaccine, made in Japan, is now available. Other types of viral encephalitis present much the same picture, and in none of them is specific treatment available. The frequency of clinical illness and the severity of meningoencephalitis vary with each specific infection.

Rift Valley fever

Long known as a serious zoonosis of domestic animals in sub-Saharan Africa, a serious human outbreak occurred in Egypt in 1977, with a strain of increased virulence. Cases with encephalitis, haemorrhagic and retinal complications occurred. The vector is *C. quinquefasciatus*.

DIRECTLY TRANSMISSIBLE HAEMORRHAGIC VIRUSES

These are diseases with a high human-to-human transmission potential, and often a relatively high case-fatality rate. These viruses are basically zoonoses which are then transmitted from person to person, mostly in the hospital situation. Although some have been presented as new diseases, it seems probable that they are really old, and that only the artificial ecology of the hospital has brought them to light. Their spectacular nature has attracted public attention disproportionate to their public health importance. Nosocomial transmission with a haemorrhagic syndrome occurs with Lassa virus and other arenaviruses, Marburg and Ebola viruses, Congo-Crimean haemorrhagic fever and Rift Valley fever.

Lassa fever

This is an arenavirus, of the group which contains lymphocytic choriomeningitis. The animal reservoir is the multimammate rat, *Mastomys natalensis*. Infected rodents are unharmed and excrete the virus for long periods in their urine. So far the disease has only been found in west Africa, but has been exported from there to Europe and north America by infected patients. Close contact with blood and body fluids of infected patients may transmit infections, so nursing, medical and laboratory staff are at special risk. The original infection is acquired from contact with rodent urine in some way. The illness was first recognized in Lassa, north-east Nigeria, in 1969, in a missionary nurse, from whom the infection spread to two colleagues. The first two nurses died. The third, from whom the virus was isolated at the Yale Arbovirus Research Unit, survived. Several outbreaks of the disease have been reported from hospitals in west Africa since then, often with high mortality, especially in pregnancy. Human infection may be frequent in endemic populations, where it causes a febrile illness with low mortality.

The onset, after an incubation period of 3–16 days, is undramatic and lacks specific features: there is fever, malaise, headache, aches and pains and sore throat. At this stage the illness resembles any minor viral infection. Symptoms continue unchanged for 3–6 days, when the patient suddenly deteriorates and is now obviously suffering from a grave illness. The main features are now:

1 High fever.
2 Severe prostration.
3 Severe sore throat with dysphagia.
4 Abdominal pain.
5 Diarrhoea and vomiting.

   The throat may now contain vesicles, exudates or ulcers and a faint rash may appear. Pleural effusions and frank haemorrhage may complicate severe cases. Death occurs on days 7–14, in a state of toxaemia and shock.

LABORATORY FINDINGS

Investigations should only be carried out in a high-security laboratory if the diagnosis is suspected. Virus isolation can only be done in a few designated centres in the world. Routine investigations are unhelpful: there is usually neutropenia and albuminuria.

MANAGEMENT

In the UK such patients have been nursed in a special isolator. In fact, the disease is much less contagious than smallpox, and there is no good evidence that transmission to hospital staff occurs if standard barrier nursing procedure is followed. These methods have been found completely adequate in hospitals in endemic areas dealing with large numbers of patients. There are conflicting reports on the value of convalescent serum in treatment, but its prophylactic use after a laboratory accident is reasonable. The mainstay of management is skilled, supportive care, but there is now unequivocal evidence that the drug ribavirin substantially increases the survival rate in severe disease. The virus may be present in the throat for 2 weeks and in the urine for 5 weeks after the onset of the illness.

CONTROL

This is mainly directed towards preventing hospital-acquired infections. The control of the disease in the population in endemic areas would require effective control of the rodent reservoir. This has never been attempted.

Bolivian and Argentinian haemorrhagic fevers

There are two other arenavirus infections causing the haemorrhagic fever syndrome. They both resemble Lassa in having a reservoir in asymptomatic rodents which excrete virus throughout their lifetime. They are:

1 Junin virus. This is responsible for seasonal rural epidemics in Argentina.

2 Machupo virus. This is responsible for sporadic outbreaks in Bolivia.

Filoviruses—Marburg virus and Ebola virus

Marburg virus is a haemorrhagic fever which was first recognized in Germany and Yugoslavia in 1967 in patients who had contact with blood or tissues from African green monkeys (*Cercopithecus aethiops*) from Uganda. Secondary infections occurred in hospital staff caring for patients. The virus may persist in the body, notably in the seminal fluid, for several months after recovery. Since then, the virus has been isolated from patients acquiring the infection in east and southern Africa. Passage beyond second-generation cases is unusual. The case-fatality rate is about 30% in primary cases.

In 1976, outbreaks of a haemorrhagic disease were reported from equatorial Sudan and Zaïre, with an overall mortality rate in excess of 50%. The clinical picture resembled Marburg disease, but the virus was antigenically distinct and named Ebola virus. There was another outbreak in the Sudan in 1979, and in all outbreaks there was strong evidence of person-to-person spread and needle passage, especially in the hospital setting, from close personal contact, especially with haemorrhaging cases. Hospitals became centres for spread of infection, and hospital staff suffered a high mortality. No animal reservoir has yet been found, but a non-human primate reservoir is strongly suspected.

CLINICAL PICTURE

After an incubation period of probably 7–10 days on average, there is an abrupt onset, with these main symptoms:

1 Severe headache.
2 High fever.
3 Prostration.
4 Generalized musculoskeletal pains.
5 Nausea and vomiting.
6 Abdominal pain.
7 Severe watery diarrhoea.
8 Conjunctivitis.
9 Stabbing chest pains.
10 Discomfort in the throat.

These symptoms commonly continue for about a week. In a large number of patients severe bleeding develops between days 5 and 7 from the gastrointestinal tract, nose, gums and vagina. Petechiae and subconjunctival haemorrhages are also very common.

DIAGNOSIS

This requires the help of a virologist with special experience. Virus can be isolated from blood and postmortem tissues and specific serology is available. The WHO will provide an expert to investigate outbreaks and assist with disease control.

Treatment is supportive only. Staff must be fully protected by standard barrier nursing techniques.

CONTROL

So far, outbreaks have terminated spontaneously, aided by the introduction of barrier nursing techniques to prevent transmission in hospital.

RABIES

This is an infection with a neurotropic rhabdovirus with a reservoir in canines and bats. Most human infections are from dog bites, the virus being excreted in the saliva. Virtually all warm-blooded animals can be infected.

Infection is almost always from inoculation, and occasionally by inhalation. Infection has also been transmitted by corneal grafts. The virus is transmitted upwards to the CNS via the nerve trunks, and causes a fatal meningoencephalitis, after proliferating in nerve cells in the brain and peripheral ganglia. Affected cells may be more irritable than normal, leading to furious rabies, or have depressed function, causing dumb rabies.

Clinical features in humans

The incubation period varies from 2 weeks to more than a year. Proximal bites and a large inoculum favour a short incubation period.

The onset is usually rapid with fever, anxiety, insomnia, and often pain or paraesthesiae at the site of the original bite. Painful spasms of the throat muscles then follow, often precipitated by attempts to swallow— hence, hydrophobia. Spasms often become more widespread, to involve the respiratory muscles. In the attacks the patient appears terrified. This may be due partly to anticipation of the painful spasm and partly to overactivity of the fear centre. Spasms may also be precipitated by air blowing on to the face—aerophobia. The patient's intellect is normal between spasms, when intelligible speech is often possible.

Death usually occurs within a week, with widespread paralysis and respiratory arrest, although victims of bat-transmitted rabies survive longer. A copious secretion of ropy saliva is characteristic, and the patient often does froth at the mouth.

Postmortem examination shows a diffuse meningoencephalitis with extensive neuronal destruction. Despite a handful of not entirely convincing reports of survival following heroic supportive treatment, death is virtually inevitable.

Diagnosis

DURING LIFE

In a proportion of cases, corneal impression smears show a positive result with fluorescein-conjugated antirabies serum, but although a positive test is diagnostic, a negative result does not exclude infection.

POSTMORTEM DIAGNOSIS

The pathologist should be warned if the diagnosis is suspected, not only so that he or she can carry out the appropriate tests, but so that he or she can take special precautions to protect other staff and him- or herself from infection.

Diagnosis can be made in two ways: examination of fixed brain tissue stained with Seller's stain for the characteristic Negri inclusion bodies, or examination of fresh brain using a fluorescent antibody technique. This is both quicker and more sensitive.

Treatment

The patient will die no matter what is done, as postmortem evidence of irreversible brain damage will eventually confirm. Intensive care can prolong life, by taking over vegetative functions, but we are unconvinced of recovery. In developing countries the option does not present itself, for resources are not available for the prolonged use of life-support systems. The most humane approach in these conditions is to relieve the agonies of the patient with effective analgesia and sedation, such as a combination of a phenothiazine, a barbiturate and heroin. In developed countries, the choice between heroic supportive measures and symptomatic treatment can only be made by the clinician in charge, after evaluation of the evidence.

PRECAUTIONS WITH RABIES PATIENTS

The patient's saliva is potentially infective. Everyone in contact with the patient should be protected as for barrier nursing, with the addition of goggles. Infection through the conjunctiva can occur, and patients often spit.

POSTEXPOSURE TREATMENT

Once symptoms of rabies have developed, the patient is doomed. But the disease can be prevented, even after exposure, because the unusually long incubation period—on average 6–12 weeks—allows effective antibody levels to be built up by active immunization during this period. If it is estimated that the incubation period is likely to be very short (as in cases of proximal bites and a large inoculum from wild-animal attack), immediate passive immunity can be conferred by an injection of hyperimmune serum. This is usually horse serum, and the usual precautions against anaphylaxis must be taken. A hyperimmune human serum is now available in the UK, free of the risks of anaphylaxis or serum sickness.

Local treatment is very effective in removing the virus from a bite, provided it is applied thoroughly and early. The WHO recommendations given later provide details of other action (Table 7.3).

TREATMENT OF RABIES

Nature of exposure	Status of biting animal irrespective of previous vaccination		Recommended treatment
	At time of exposure	During 10 days	
Contact, but no lesions; indirect contact; no contact	Rabid	–	None
Licks of the skin; scratches or abrasions; minor bites (covered areas of arms, trunk and legs)	Suspected as rabid	Healthy	Start vaccine. Stop treatment if animal remains healthy for 5 days
	Rabid; wild animal, or animal unavailable for observation	Rabid	Start vaccine; administer serum upon positive diagnosis and complete the course of vaccine. Serum + vaccine
Licks of mucosa; major bites (multiple or on face, head, finger or neck)	Suspect rabid domestic or wild animal, or animal unavailable for observation		Serum + vaccine. Stop treatment if animal remains healthy for 5 days

Table 7.3 Action to be taken after rabies exposure.

ANTIRABIES VACCINES

In developing countries the most commonly used vaccine is Semple vaccine, prepared from animal brain tissue infected with fixed virus and inactivated with phenol. This is usually effectively antigenic, but may cause postvaccinial encephalitis in as many as 1 in 250 vaccinees. Corticosteroid treatment reduces mortality and morbidity if given as soon as symptoms of encephalitis or myelitis develop.

The new vaccines grown in human diploid cells (HDCV) are potent, and apparently free of the danger of producing neuroanaphylactic accidents. But they are very expensive (£18 at 1994 prices for one dose of 1 ml). HDCV effectiveness is greatly enhanced by giving it intradermally, when 0.1 ml is as effective as 1 ml subcutaneously. It has been shown that 8 intradermal doses of 0.1 ml given into different sites in one session can produce protective levels of antibody within a week. This offers enormous advantages compared with the traditional need for prolonged courses, both in cost and patient compliance. Booster doses of 0.1 ml intradermally at 7 and 28 days maintain the antibody response.

Other equally potent, inactivated and purified vaccines are purified chick embryo cell vaccine (PCECV), purified vero cell vaccine (PVCV) and purified duck embryo vaccine. These newer vaccines are one-third the cost of HDCV and lend themselves to economical multisite use.

THE BEST CHOICE FOR DEVELOPING COUNTRIES

A study in Thailand showed successful results in 328 cases, many at high risk, given 0.1 ml HDCV intradermally on days 0, 3, 7 and 14 only. Antibody levels were good.

PRE-EXPOSURE VACCINATION

Three spaced doses of HDCV (consult package insert) give good levels of immunity which should be adequately protective for those at special risk, such as veterinary surgeons working in endemic areas. What the manufacturers do not say is that 0.1 ml intradermally seems to be as effective as 1 ml subcutaneously. Once the freeze-dried vaccine is reconstituted, it retains its potency for several days if kept at 4°C. The normal preexposure regimen is 0.1 ml intradermally at days 0, 7 and 28. Failure of the regimen has been recorded, and might be related to concomitant chloroquine administration.

Rabies in the dog

This is usually furious: the dog rushes round emitting a high-pitched bark, and biting not only people but objects. An unprovoked attack on the owner is a typical first sign. The animal may paw at its mouth, as if trying to dislodge a foreign body, and salivate excessively. Ten days after it becomes infective, it will be dead. This fact is made use of when a dog which has bitten someone can be impounded. If the dog is alive 10 days after the bite, it could not have been infective when it bit. There are disquieting reports that some asymptomatic dogs can excrete virus for long periods.

Dogs can be given a high degree of protection against rabies by an attentuated live vaccine.

Rabies in other animals

Dogs, foxes, wolves and jackals are the major reservoirs of rabies in most parts of the world, and cats can also transmit infection, usually themselves suffering from a paralytic illness typical of 'dumb rabies'. In south America and the Caribbean, bats are very important and cause enormous economic losses due to loss of cattle from paralytic rabies. Bats can also cause human disease, not only from their bites, but from the inhalation of infected guano by speleologists exploring bat caves, mainly in the Americas.

Prevention of rabies

The most useful measure is usually control of domestic dogs, and this merely requires an effectively enforced vaccination programme combined with regular dog-catching. Unvaccinated dogs (those not wearing a collar with a vaccination tag) should be regularly caught and destroyed.

The best results are obtained with professional dog-catchers using bait, who operate from a vehicle, and take the animals back to their headquarters for destruction. Attempts to control dogs by soldiers armed with assault rifles are invariably unsuccessful and dangerous.

Size of the problem

The problem is enormous, and most developing countries do not do much to prevent it. In some parts of India, 1 in 500 hospital admissions is due to rabies, almost all due to dog bites. The country has more than 1500 rabies vaccination centres and manufactures 25 800 l of vaccine a year.

Of all dogs caught in Bangkok, 1 or 2% are rabid, and there are 300 deaths a year in that city from rabies.

WHO guide for postexposure treatment

These recommendations are intended only as a guide. It is recognized that in special situations modifications of the procedures may be warranted. Such situations include exposure of young children and other circumstances where a reliable history cannot be obtained, particularly in areas where rabies is known to be endemic, even though the animal is considered to be healthy at the time of exposure. Such cases justify immediate treatment, but of a modified nature, e.g. local treatment of the wound as described below, followed by administration of a single dose of serum or three daily doses of vaccine; provided that the animal stays healthy for 10 days following exposure, no further vaccine need be given. Modification of the recommended procedures would also be indicated in a rabies-free area where animal bites are frequently encountered. In areas where rabies is endemic, adequate laboratory and field experience indicating no infection in the species involved may justify local health authorities in recommending no specific antirabies treatment.

Practice varies concerning the volume of vaccine per dose and the number of doses recommended in a given situation. In general, the equivalent of 2 ml of a 5% brain-tissue vaccine, or the dose recommended by the producer of a particular vaccine, should be given daily for 14 consecutive days. To ensure the production and maintenance of high levels of serum-neutralizing antibodies, booster doses should be given at 10, 20 and 90 days following the last daily dose of vaccine in *all* cases.

Combined serum-vaccine treatment is considered by the WHO Committee as the best specific treatment available for the postexposure prophylaxis of rabies in humans. Experience indicates that vaccine alone is sufficient for minor exposures. Serum should be given in a single dose of 40 IU/kg of body weight for heterologous serum and 20 IU/kg of body weight for human antirabies immunoglobulin; the first dose of vaccine is inoculated at the same time as the serum, but at another site.

Treatment should be started as early as possible after exposure but in no case should it be denied to exposed persons whatever time has elapsed.

Where antirabies serum is not available, full vaccine therapy, including three booster inoculations, should be administered.

LOCAL TREATMENT OF WOUNDS INVOLVING POSSIBLE EXPOSURE TO RABIES RECOMMENDED IN ALL EXPOSURES

I First-aid treatment. Since elimination of rabies virus at the site of infection by chemical or physical means is the most effective mechanism of protection, immediate washing and flushing with soap and water, detergent or water alone is imperative. Then apply 40–70% alcohol, tincture or aqueous solutions of iodine, or 0.1% quaternary ammonium compounds.*

2 Treatment by or under direction of a physician:

(a) Treat as in (1) then:

(b) Apply antirabies serum by careful instillation in the depth of the wound and by infiltration around the wound.

(c) Postpone suturing of wound; if suturing is necessary, use antiserum locally, as stated above.

(d) Where indicated, institute antitetanus procedures and administer antibiotics and drugs to control infections other than rabies.

FURTHER READING

World Health Organization (1985) *Arthropod Borne and Rodent Borne Viral Disease.* Technical Report Series 719. World Health Organization, Geneva.

In: *Modern Vaccines: Current Practice and New Approaches.* Rabies pp 113–20. Ed Adviser E Richard Moxon: Edward Arnold.

* Where soap has been used to clean wounds, all traces of it should be removed before the application of quaternary ammonium compounds because soap neutralizes the activity of such compounds.

Typhoid and paratyphoid fevers (enteric fevers)

By the terms typhoid and paratyphoid fevers I mean those illnesses caused by *Salmonella typhi* and *S. paratyphi* A, B and C. They all often cause a systemic, septicaemic illness, but so sometimes do the many zoonotic salmonellas which more usually cause *Salmonella* food poisoning.

Typhoid and paratyphoid organisms are cosmopolitan in distribution, but are commonest where standards of personal and environmental hygiene are low. Only to this extent are these diseases tropical.

Organisms

The causal organisms are all Gram-negative bacilli with flagella. All possess somatic (O) and flagellar (H) antigens. *S. typhi* and *S. paratyphi* C sometimes possess a surface (Vi) antigen that coats the O antigen and potentially protects it from antibody attack.

S. typhi and *S. paratyphi* A and B usually infect only humans. *S. paratyphi* C may affect a variety of animals also.

Mode of infection

This is virtually always by ingestion. Infection may be transmitted in water (mainly *S. typhi*) and food, and is largely dose-related. Ingestion of a fairly small dose of *S. typhi*, such as 10^5 organisms, may cause a relatively low attack rate with a fairly long incubation period. But increasing the infecting dose to 10^9 organisms raises the attack rate to 95% and greatly shortens the incubation period. High gastric acidity opposes infection.

All these organisms can multiply in suitable foods maintained at a favourable temperature, and so greatly enhance the efficiency of human food-handlers in transmitting the infection. The most important reservoirs of infection are asymptomatic human carriers.

TYPHOID FEVER

After ingestion, the organisms attach to the small intestinal mucosa, penetrate it. and are transported by the lymphatics to mesenteric lymph glands. There they multiply, and enter the blood stream via the thoracic duct. The main location of bacilli is inside macrophages. From this bacteraemia, which corresponds to the end of the incubation period, organisms are carried to the bone marrow, spleen, liver and gallbladder.

There is now a secondary invasion of the bowel via the infected bile. Organisms multiply in macrophages, and pathological changes are greatest where macrophages are present in large numbers, such as in the intestinal lymph follicles. The largest of these are Peyer's patches in the ileum.

There is now a strong inflammatory response with infiltration by inflammatory cells (macrophages and lymphocytes), and the Peyer's patches become hyperplastic. If inflammation does not resolve, necrosis occurs within 7–10 days and the patches ulcerate. Involvement of blood vessels may lead to bleeding and, if the whole thickness of the bowel is involved, perforation follows.

Elsewhere in the body, foci of inflammation with macrophages and lymphocytes, so-called typhoid nodules, are scattered in various organs, especially the liver, spleen, marrow and lymph glands.

More diffuse organ involvement also occurs, such as cloudy swelling of hepatocytes, necrosis, degeneration and fatty infiltration of the myocardium, degenerative changes in kidney tubules and interstitial pneumonitis. Late in the disease there may be abscess formation, in which polymorphonuclear cells now predominate, most often affecting bone, brain, liver or spleen.

The natural course of the disease is very variable. In a classical case, fever has returned to normal at the end of the third week and repair processes then begin. In some cases fever may continue for many weeks, and others are abortive, with a brief and unspectacular course.

Death most commonly results from perforation, haemorrhage or toxaemia, occasionally from other complications such as meningitis.

No doubt much of the pathology is due to the obvious local inflammatory response to the bacilli, as in the gut. But serious disease of brain, lung and kidneys is not usually accompanied by typhoid nodule formation, and the assumption is that some unidentified toxin must be the cause. The well-studied S. typhi endotoxin does not seem to be the culprit.

Clinical picture

The incubation period is about 14 days on average, but can vary from less than a week to more than 3 weeks. The only almost constant symptoms are fever and headache. The untreated illness normally runs its course in about 3 weeks but can extend to months in exceptional cases.

CLASSICAL PATTERN OF FEVER

The onset is usually gradual, and rigors are unusual. Fever increases day by day in the first week, often with an evening rise. A remittent fever then continues for another week or more, then falls by lysis in the third week.

The pulse rate usually is relatively slow compared with the fever, and may not reach 100 beats/min even when the temperature is 40°C.

OTHER SYMPTOMS IN UNCOMPLICATED CASES

Patients with typhoid usually feel very unwell in general, with malaise, generalized aches and pains, and anorexia. The following symptoms are also common:

1 Abdominal pain or discomfort.
2 Constipation.
3 Diarrhoea.
4 Deafness.
5 Cough.

PHYSICAL SIGNS

These depend not only on the severity of the illness but on the length of time the patient has been ill.

In patients who seek medical aid early, there has usually been no significant dehydration from diarrhoea, and the patient often looks relatively well and is mentally alert. In contrast, the patient who presents after 2 weeks of illness is often very toxic, mentally stuporose and gravely dehydrated. The commonest signs are:

1 Fever.
2 A disproportionately slow pulse.
3 Hepatomegaly.
4 Splenomegaly, often tender.
5 Mental changes.
6 Signs of bronchitis.
7 Rose spots.
8 Deafness.
9 Meningism.

Rose spots can only be seen in fair-skinned patients. They are found from day 7 onwards and take the form of pink macules, usually scanty, mainly on the trunk, They fade on pressure from a glass slide.

Enlargement of the liver and spleen occurs after only a few days of illness, but may be delayed.

COMPLICATIONS

These may develop as the illness progresses, and they may follow a clinically mild attack. So the clinician must remember that typhoid patients may present with the complication itself, rather than with the symptoms of typhoid fever. These patients are often difficult diagnostic problems.

1 *Perforation.* This typically occurs in the third week. Toxic patients show few signs of peritonitis, except for abdominal distension, increasing toxaemia and a rising pulse. Surgery is nowadays considered to give a better chance of survival than conservative management, and excision or segmental resection is safer than simple suturing, for the gut wall immediately surrounding the perforation may be too friable to hold sutures.

2 *Haemorrhage.* This is also typically a complication of the third week. There may be massive bleeding or repeated small bleeds. Surgery is seldom needed provided blood transfusion is available.

3 *Haemolytic anaemia.* This is common in patients with G6PD deficiency and typhoid depresses G6PD levels in normal as well as in deficient patients.

4 *Typhoid lobar pneumonia.* This is a rare complication of the second and third week. Rusty sputum is not produced.

5 *Meningitis.* This may be the only obvious manifestation of typhoid, when it resembles any other pyogenic meningitis.

6 *Renal disease.* This may present as renal failure or an acute nephrotic syndrome, and is probably an immune-complex nephritis. Recovery after successful chemotherapy is usual.

7 *Typhoid abscess.* This is a late complication that can occur almost anywhere, especially in the spleen, liver, brain, breast and skeletal system.

8 *Skeletal complications.* These are mainly suppurative arthritis and osteomyelitis. Both may be greatly delayed in onset. Zenker's degeneration of muscle or polymyositis may occur.

9 *Other complications or sequelae.* These include suppurative parotitis, acute cholecystitis, deep venous thrombosis and the Guillain–Barré syndrome.

Diagnosis

CULTURE

Culture of the organism is the mainstay of diagnosis. Unfortunately, this technology is often lacking in those hospitals in developing countries that most need it.

Blood culture is usually regarded as the most useful technique in the first week, but may be positive at any later stage. A significantly higher rate of positivity occurs with marrow cultures (95 versus 43%) even if chemotherapy has been started.

Stool culture often becomes positive in the second week, or earlier if the patient has diarrhoea. Recently a string capsule used to sample duodenal contents has been found to give a much better culture-positive rate (86 versus 42%) than blood culture.

Urine culture becomes positive in about 25% of cases after the second week, but its main use is in the detection of urinary carriers.

Other materials, such as aspirates from rose spots, CSF or pus from abscesses, also yield positive culture results at times.

SERODIAGNOSIS

The most widely used test is the Widal test, which measures agglutinating antibodies to the somatic (O) and flagellar (H) antigens.

The test is essentially non-specific, because numerous non-typhoid salmonellae share O and H antigens with S. typhi. Its usefulness is greatly diminished by three other factors:

1 H antibody titres remain high for a long time after typhoid immunization.

2 In typhoid patients titres often rise before the clinical onset, making it very difficult to demonstrate the diagnostic fourfold rise between initial and subsequent specimens.

3 A significant number of culture-positive patients develop no rise in titre at all.

But if the test is interpreted intelligently, bearing all these facts in mind, a significant number of patients will be correctly diagnosed by the Widal test, when all other methods have failed.

Tests for detecting S. typhi antigen in serum, potentially far more useful, are still in their infancy.

OTHER LABORATORY FINDINGS

The WBC is usually within the normal range, as is the differential count, but there may be leucopenia or leucocytosis, and relative lymphocytosis is common.

Biochemical tests usually show only minor changes, such as slight elevation of transaminases and bilirubin. A considerable elevation of indirect bilirubin is often associated with haemolytic anaemia in patients with G6PD deficiency and in children. Huckstep, who accumulated great experience of typhoid in east Africa, claimed that the diazo test in urine was an extremely useful diagnostic test in typhoid.

In severe cases, albuminuria is almost invariable, and there may be evidence of DIC.

Treatment

Good nursing care is essential, for patients are often desperately ill and mentally uncooperative on first admission, and need attention to all their bodily needs. A high quality of good supportive medical care—as in the maintenance of fluid and electrolyte balance—is also vital to achieve good survival rates. But the mainstay of treatment is effective antimicrobial chemotherapy.

CHEMOTHERAPY

Chloramphenicol used to be acknowledged everywhere as the drug of choice. But in recent years resistance has been reported and in some places resistance is now a major problem: in Bangkok almost half of all strains are chloramphenicol-resistant and significant numbers of cases are also resistant to ampicillin and co-trimoxazole. Resistance to all three drugs may occur in 40% of all cases in parts of the Indian subcontinent.

The newer quinolone antibiotics such as ciprofloxacin, ofloxacin and norfloxacin are usually effective, but even with them, resistance has been reported. A 5–7-day course of a third-generation cephalosporin such as ceftriaxone is effective in multidrug-resistant strains, but the relapse rate is uncertain.

Chloramphenicol

The drug is bacteriostatic only. A fairly prolonged course must be given to prevent relapse, such as a total of 14 days, or 12 days after fever has abated.

Successful regimens mostly lie in the range of 1 g 6-hourly to 0.5 g 4-hourly until the patient is afebrile, followed by 500 mg 6-hourly for 10–12 days. It commonly takes 48 h before the fever shows a response, and 5 days or more until the patient becomes completely afebrile in severe cases.

A Herxheimer-type reaction is sometimes seen early in treatment, and should be treated with steroids.

Chloramphenicol remains the drug of choice in Africa and Papua New Guinea.

Amoxycillin

This is more expensive than chloramphenicol, but at least as effective if given in high doses, and essential where there is chloramphenicol resistance. Doses in the range 500 mg–1 g 6-hourly for 14–21 days are usual. Ampicillin is inferior to chloramphenicol.

Co-trimoxazole

The dose is two tablets 8-hourly until afebrile, then two tablets 12-hourly for 10 days. The clinical response is at least as rapid as with chloramphenicol.

USE OF STEROIDS WITH CHEMOTHERAPY

Intensely toxic patients show a dramatic improvement in general condition when steroid therapy is given, and survival may be improved. Perforation has not proved a problem, but steroids should be avoided in the third week of illness.

PROBLEM OF RELAPSE

Relapses after chemotherapy occur in a variable proportion of patients (2–10%), are usually rather less severe than the initial illness, and respond to the same chemotherapy.

CARRIER STATE

This commonly persists for some months into convalescence, and when it terminates spontaneously such patients are called convalescent carriers. They are an obvious source of infection to others, but even more important are chronic carriers (1–3% of cases) in which a persisting focus of infection smoulders on in the gallbladder (faecal carriers) or urinary tract (urinary carriers). In most endemic areas few carriers are identified, because culture facilities do not exist. Persistent elevation of Vi antibodies often accompanies the carrier state.

Chronic carriers may, when employed as food-handlers, cause hundreds of cases and many deaths. The excretion of organisms by carriers is variable and erratic. Chronic faecal carriers usually have chronic cholecystitis with or without gallstones, and there are often pathological abnormalities in the urinary tract, including *Schistosoma haematobium* infection in chronic urinary carriers. *S. mansoni* may be associated with a relapsing non-typhoid *Salmonella* septicaemia.

TREATMENT OF CHRONIC CARRIERS

Co-trimoxazole, two tablets twice a day for 3 months, is the most cost-effective treatment. Ampicillin and amoxycillin in high dosage for the same length of time, combined with probenecid, may also be effective. But faecal carriers with gallstones only respond temporarily to chemotherapy and cholecystectomy is needed to terminate the carrier state in such cases.

Recently, norfloxacin at a dose of 400 mg twice daily for 28 days was shown to cure 11 of 12 carriers.

If the patient is intelligent and conscientious, and not a food-handler, the carrier state need not be treated at all, for the fastidious maintenance of high standards of environmental and personal hygiene will prevent transmission of the infection to others.

TYPHOID VACCINE

A monovalent vaccine using killed S. typhi organisms is most widely used. The old typhoid and paratyphoid vaccine A and B (TAB) contained paratyphoid organisms also but was never proved to provide worthwhile immunity to the paratyphoids, and often produced more severe reactions than the monovalent vaccine. A modern purified surface antigen vaccine is now in use, but is very expensive (over £10 a dose). The live attenuated oral vaccine requires three doses at a cost of about £15 to provide 1 year's immunity.

The degree of protection given by the vaccine is about 90%. If typhoid does develop in a vaccinated subject, it is no less severe than in the unvaccinated.

PARATYPHOID A AND B

These usually infect via contaminated foods in which the organisms have multiplied. For this reason, diarrhoea and vomiting may precede septicaemia. Many mild cases occur. Treatment is as for typhoid.

PARATYPHOID C

This commonly produces septicaemia without involvement of the gut, and abscess formation is common.

FURTHER READING

Gupta, A. (1994) Multidrug-resistant typhoid fever in children: epidemiology and therapeutic approach. *Paediatr Infect Dis J* 13, 134–40.

Tuberculosis in the tropics*

HISTORICAL ASPECTS

During the course of the 20th century the dramatic decline in the incidence of tuberculosis (TB) in industrialized countries was primarily due to improved socioeconomic conditions.

The chances of achieving successful cure in an individual with active TB remained low until the discovery of effective antituberculous drugs: streptomycin in 1946, isoniazid in 1952, pyrazinamide in 1954 and rifampicin in 1970.

EPIDEMIOLOGY

Size of the problem

There are estimated to be 8–10 million new TB cases diagnosed in the world each year and the disease is responsible for the death of 2–3 million people annually. The bulk of this morbidity and mortality occurs in tropical and subtropical regions where infection rates approach 2–3% per year of life. The incidence of TB is actually increasing, so that there are more new cases now than 20 years ago. The single most important factor in the resurgence of TB is the AIDS pandemic, as infection with HIV greatly increases the risk of TB infection and disease.

Transmission

Infection with the tubercle bacillus is by inhalation of droplet nuclei which have been coughed up by someone with active pulmonary disease. This is the only way that infection with *Mycobacterium tuberculosis* can occur. A sputum-smear-positive individual is 10 times more infectious than someone who is smear-negative, so the emphasis of TB control in the

* This chapter is based on lectures originally given by Professor A.D. Harris and Dr R. Fox.

tropics should remain the detection of these infectious individuals, even in the presence of high HIV seroprevalence.

Risk of progression to disease

Up to 15% of infected people develop the disease–5% within 2 years of infection, 5% within 5 years and the rest during the remainder of their lifetime.

Factors which influence infection and progression include:

1 Intensity of exposure: overcrowding.
2 Age: very young and very old.
3 Genetic susceptibility: weak association with human leucocyte antigen (HLA) type.
4 Immunosuppression: HIV, malnutrition, transplant recipients.
5 Trauma: TB osteomyelitis.
6 Miscellaneous: alcohol.

PATHOGENESIS AND CLINICAL SYNDROMES (FIG. 9.1)

The granulomatous reaction in the terminal air spaces with enlarged regional lymph nodes occurring 3–8 weeks after infection is the primary complex and is usually asymptomatic and heals spontaneously. Some people will present with manifestations of tuberculin reactivity, such as

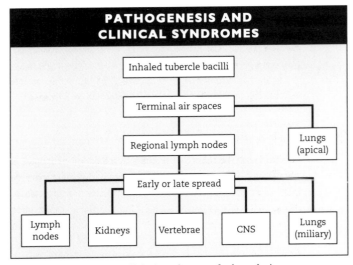

Fig. 9.1 Pathogenesis and clinical syndromes of tuberculosis.

erythema nodosum, phlyctenular conjunctivitis or pleural effusion. The lymphadenopathy may cause lobar collapse due to external compression of a bronchus. Early dissemination occurs most commonly in children presenting as miliary TB or TB meningitis, although the disease can affect any organ system. Classical cavitating TB affecting one or both lung apices is due to reactivation of dormant organisms at the site of primary infection and is the commonest presentation in adults.

PULMONARY TUBERCULOSIS

Pulmonary TB usually presents with chronic cough with blood-stained sputum, weight loss, fevers and sweats. Clinical examination is often unhelpful but most patients will be malnourished and some will have finger clubbing or focal signs in the chest. The single most important investigation is a sputum smear to look for acid-fast bacilli (AFB). The best-quality sputum samples are obtained in the morning and should be stained using the Ziehl–Neelsen method and examined under oil immersion using a light microscope. Fluorescent microscopy is quicker but expensive and is appropriate only in centres where large numbers of samples are processed. Culturing the sputum of smear-positive individuals for mycobacteria is not routinely required as it is expensive and plays little part in controlling the transmission of the disease.

The chest X-ray changes of pulmonary TB are most often in the apices with consolidation and cavitation. The differential diagnosis includes lung abscess, pulmonary paragonimiasis, nocardiosis, actinomycosis, histoplasmosis, melioidosis and bronchial carcinoma and, although these are all much less common than TB, one must be wary of diagnosing the disease on the basis of a chest X-ray alone.

Tuberculin test

Testing for cutaneous reactivity to tuberculin either by the Mantoux or Heaf test is of limited value in the diagnosis of active TB. A strongly positive test (>10 mm induration) suggests active disease or recent infection; however, a weakly positive (<10 mm induration) or even a completely non-reactive test does not exclude active TB. Many HIV-positive patients with TB will have a false-negative tuberculin test.

Serodiagnosis and DNA techniques

Enormous research effort has gone into finding alternative methods for diagnosing TB, but for pulmonary disease none have yet to supersede sputum microscopy. Serodiagnosis using monoclonal antibodies or PCR to detect mycobacterial DNA may find a place in extrapulmonary or paucibacillary disease and DNA fingerprinting is proving to be a superb epidemiological tool with which to study the transmission of *Mycobacterium tuberculosis* in human populations. However, examining sputum smears will remain the cornerstone of TB control in the tropics for the foreseeable future.

TUBERCULOUS LYMPHADENOPATHY

The differential diagnosis includes the persistent generalized lymphadenopathy associated with HIV infection, Kaposi's sarcoma and lymphoma. Confirming the diagnosis can be difficult if facilities are limited but HIV-negative cases will have a strongly positive tuberculin reaction. Lymph-node aspiration with staining for AFB should be performed and, if possible, lymph-node biopsy for histology and culture.

MILIARY HAEMATOGENOUS DISSEMINATION

Miliary TB is to due to blood spread of tubercle bacilli either following a recent primary infection or following reactivation of an old lesion. Symptoms tend to be non-specific, consisting of fever, malaise and weight loss. Clinical examination may reveal hepatosplenomegaly and ophthalmoscopy may show choroidal tubercles. Sputum smears should be performed but are often negative, although the chest X-ray may show characteristic miliary shadowing. The tuberculin test is very unreliable and is often negative. Most patients are anaemic with a normal or low peripheral WBC and a high ESR. Tests which require more sophisticated facilities but which may be helpful if available include urine culture, bone-marrow smear and culture and liver biopsy.

Miliary TB is an important cause of pyrexia of unknown origin and a therapeutic trial of antituberculous drugs may be required in cases where the diagnosis cannot be confirmed. I would recommend isoniazid in combination with either pyrazinamide or ethambutol, which is a highly specific antituberculous combination, and patients with the disease should respond within 2–4 weeks, when rifampicin can be added.

TUBERCULOUS MENINGITIS

The clinical presentation of tuberculous meningitis is very variable, ranging from headache, fever and behavioural changes to convulsions and deep coma. Arachnoiditis affecting the brainstem may cause cranial nerve palsies or a polyradiculopathy if it involves the spinal nerve roots. Most patients will have a stiff neck and lumbar puncture is mandatory. Typical CSF findings are a lymphocyte pleocytosis, elevated protein and low glucose. Staining the CSF for AFB will confirm the diagnosis in only 25% of cases and polymorphs may outnumber lymphocytes early in the disease. A high index of suspicion is therefore required and treatment may have to be commenced without a definitive diagnosis. The main differential diagnoses are partially treated bacterial meningitis, viral meningitis and cryptococcal meningitis.

BONE AND JOINT TUBERCULOSIS

TB of the spine is the commonest cause of paraplegia in the tropics. It affects the thoracic more often than the lumbar spine and causes a characteristic gibbus deformity due to collapsed vertebral bodies.

Joint TB is usually monoarticular, affecting the hip or knee, and should be suspected in any patient with a chronic painful large joint. Diagnosis is usually obvious on X-ray.

GENITOURINARY TUBERCULOSIS

Renal TB should be suspected in any patient with chronic dysuria and frequency of micturition who fails to respond to standard antibiotics. The urine is abnormal in 90% of cases, namely a sterile pyuria often with microscopic haematuria. Microscopy for AFB is unreliable and usually negative, so early-morning urine should be cultured for TB if facilities are available. Plain abdominal X-ray may show renal or ureteric calcification and intravenous pyelography is usually abnormal with caliceal irregularities or hydronephrosis.

TB of the male genital tract presents with scrotal masses which may discharge on to the surface. It often coexists with renal TB. In females the disease presents as infertility or with symptoms of pelvic inflammatory disease.

GASTROINTESTINAL AND ABDOMINAL TUBERCULOSIS

TB can affect the gastrointestinal tract anywhere from mouth to anus and can be very difficult to diagnose. It most often causes disease in the ileocaecal region and presents with abdominal pain, fever and diarrhoea. Stricture formation is common and may cause intermittent subacute obstruction. A stricture may be visible on plain abdominal X-ray but barium studies should be performed if available. Laparoscopy is also very useful by allowing inspection of the abdominal cavity and biopsy of suspicious areas under direct vision.

TUBERCULOSIS AND HIV

There is a major interaction between M. tuberculosis and HIV, causing a huge public health problem in poor tropical countries where infection with both agents is common. The most obvious manifestation of this is the rising incidence of all forms of TB in tropical Africa, so that many more people are requiring antituberculous treatment. Up to half of these cases are HIV-positive.

The clinical implications are both diagnostic and therapeutic. Many HIV-positive patients with TB will present in the same way as those who are HIV-negative, but there is a tendency for more sputum-smear-negative disease with less cavitation, more lymphatic disease and more extrapulmonary disease, especially pleural and pericardial. Active TB can occur at any stage of HIV infection, not only in the profoundly immunosuppressed.

The principles of treatment are similiar in HIV-positive patients and the initial response is often very good. Overall mortality is increased, but this is due to infections other than TB, such as severe pneumococcal or Salmonella bacteraemias. The recurrence rate following treatment may be increased; this is often due to reinfection rather than true relapse.

There is an increased incidence of severe cutaneous hypersensitivity reactions to drugs—especially with thiacetazone—which can be life-threatening.

The role of primary and secondary antituberculous chemo-prophylaxis in HIV positive individuals is an area of controversy and intense debate at the present time.

CHEMOTHERAPY FOR TUBERCULOSIS

Basic principles

Wherever possible, use a regimen shown to be effective in controlled trials and emphasize the importance of continuing treatment for the prescribed duration. Infectiousness disappears within 2 weeks, symptoms disappear within 4 weeks and sputum should be smear-negative within 2–3 months. Compliance will be impossible to achieve if the drugs are not supplied free of charge. If there is a national TB control programme in operation you must notify the patient to the programme.

The essential antituberculous drugs, their usual abbreviations and dosage are given in Table 9.1.

Thiacetazone and ethambutol are bacteriostatic drugs which are used to prevent emergence of resistance to the others, which are bactericidal.

ANTITUBERCULOUS DRUGS

Drug	Abbreviation	Daily		Intermittent	
		mg/kg	mg-max	mg/kg	mg-max
Isoniazid	H	8*	300	15	900
Rifampicin	R	10	600	10	600
Streptomycin	S	15–20	1000	15–20	1000
Pyrazinamide	Z	30	2000	50	3500
Thiacetazone	T	2.5	150	–	–
Ethambutol	E†	15	1200	40	2000

* 10 mg/kg in children up to 10 years old.
† Do not give ethambutol to children.

Table 9.1 The essential antituberculous drugs, together with their usual abbreviations and dosage.

Single preparations

ISONIAZID
Tablets 50, 100 and 300 mg.

RIFAMPICIN
Tablets 150 and 300 mg.
Adults >50 kg = 600 mg once daily.
Adults <50 kg = 450 mg once daily.

PYRAZINAMIDE
Tablets 500 mg.
Adults >50 kg = 2 g once daily.
Adults <50 kg = 1.5 g once daily.

THIACETAZONE
Tablets 50 and 150 mg.

ETHAMBUTOL
Tablets 100 and 400 mg.

Combined preparations

Isoniazid combined with rifampicin (Rifinah, Rimactazid).
Isoniazid combined with ethambutol (Mynah).
Isoniazid combined with thiacetazone (Thiazina).

FAILURE AND TOXICITY RATES

Duration	Regimen	Failures (%)	Toxicity (%)
12 months	1STH/11TH	5–10	4
12 months	2STH/10TH	5–10	4
12 months	1STH/11SH$_2$	5–10	4
12 months	1SH/11SH$_2$	5–10	4
8 months	2SRHZ/6TH	0–3	1
8 months	2SRHZ/6EH	0–3	1
6 months	2SRHZ/4RH	0–2	1
6 months	2ERHZ/4RH	0–2	1
6 months	2SRHZ/4RH$_2$	0–4	1

*The number before the first letter of the phase of the regimen is the duration in months of that phase. The number in subscript after the last letter is the number of doses per week in the maintenence phase of an intermittent regimen.

Table 9.2 Duration, failure rate and toxicity levels for antituberculous drugs. See Table 9.1 for drug abbreviations.

Drug regimens

Regimens are an initial intensive phase followed by a longer maintenance phase. The choice of combination and the total duration of therapy depend essentially on availability. The longer non-rifampicin-containing regimens are sometimes called standard chemotherapy and the shorter-duration combinations, which include rifampicin, are called short-course chemotherapy. The latter should be used wherever possible as cure rates are much higher and so more cost-effective. Isoniazid is always given for the duration of therapy and regimens which do not include rifampicin require 12 months in total. Those regimens which include rifampicin in the intensive phase but not the maintenance phase require 8 months' duration. Regimens of 6 months' duration must include pyrazinamide for the first 2 months and rifampicin for the whole 6 months. Thiacetazone should be avoided unless the patient is known to be HIV-negative.

In extrapulmonary disease the duration of therapy is the same, except in meningitis when the maintenance phase has to be prolonged to 18 months.

The duration of therapy, failure rate and toxicity levels are given in Table 9.2 for the antituberculous drugs described above.

Toxicity of regimens containing thiacetazone is much higher in those who are HIV-positive.

Treatment failure or relapse

The above regimens are suitable for new cases. Management of treatment failures, relapses and poor compliance depend on the circumstances.

1 Patients who default during the intensive phase should be retreated from the beginning with the same drugs.

2 Patients who default for less than 1 month during the maintenance phase should continue and complete the maintenance phase.

3 Patients who default for more than 1 month or are lost during the maintenance phase should be retreated from the beginning of the intensive phase with the same regimen.

4 Patients who fail to respond to adequate therapy or relapse following a full course of adequate therapy should be assumed to harbour resistant organisms. This may also occur if compliance is erratic throughout treatment. A regimen suitable for such patients would be 3SRHZE/6HRZE (see Table 9.1 for drug abbreviations). Sputum should be sent to the regional laboratory for culture and sensitivity testing and the patient admitted to hospital for at least the whole of the intensive phase.

Adverse effects of essential antituberculous drugs

ISONIAZID

1 Hepatitis—very rare under the age of 30 years; it occurs in 2% of those over 50 years.

2 Peripheral neuropathy—prevent with pyridoxine 10 mg once daily.

3 Hypersensitivity reactions.

RIFAMPICIN

1 Gastrointestinal effects; rifampicin colours urine orange-red.

2 Hepatotoxic; it potentiates the hepatotoxic effects of isoniazid.

3 Immunological, hypersensitivity reactions and 'flu syndrome' with intermittent use.

4 Rarely, renal damage, thrombocytopenia and haemolysis.

5 Rarely, pseudoadrenal crisis.

6 It induces liver enzymes.

STREPTOMYCIN

1 Hypersensitivity reactions.

2 Eighth nerve damage—vestibular damage is more frequent than auditory.

3 Nephrotoxic.
4 Total cumulative dose should not exceed 100 g.

PYRAZINAMIDE

1 Hepatotoxic.
2 Arthralgia – related to hyperuricaemia.

THIACETAZONE

1 Generalized cutaneous reactions.
2 Gastrointestinal effects.
3 Adverse reactions are common with HIV.

ETHAMBUTOL

1 Retrobulbar neuritis – very rare if dose is 15 mg/kg.

Hypersensitivity reactions and hepatitis

Hypersensitivity reactions present with rash and/or fever usually during the first 1–2 months of commencing therapy. Mild reactions can be treated with oral antihistamines but severe reactions may require corticosteroids and interrupting therapy. Attempt to identify the drug or drugs responsible and resume adequate treatment as soon as possible.

Hepatitis is often due to the rifampicin–isoniazid combination. All drugs should be stopped and the patient sequentially rechallenged when liver function has recovered. Mild cases may tolerate the same regimen. Table 9.3 shows the challenge doses used with antituberculous drugs.

CHALLENGE DOSES		
Challenge doses	**Day 1**	**Day 2**
Isoniazid	50 mg	150 mg
Rifampicin	75 mg	300 mg
Pyrazinamide	100 mg	250 mg
Ethambutol	100 mg	400 mg
Thiacetazone	25 mg	50 mg
Streptomycin	125 mg	500 mg
For severe reactions use 1/10th full dose on day 1. Avoid rechallenge with thiacetazone in patients who are HIV-positive.		

Table 9.3 Challenge doses used when reintroducing antituberculous drugs after hypersensitivity reactions.

Use of corticosteroids

Consider using for any severe drug hypersensitivity reactions and all seriously ill patients with any form of TB. To prevent formation of fibrous tissue, always use in patients with meningitis, pericarditis or ureteric disease. If steroids are given in conjunction with rifampicin, the dose should be 60 mg or equivalent of prednisolone per day, because of the enzyme-inducing effect of rifampicin. The same effect on liver enzymes may make the contraceptive pill ineffective.

BOX 9.1: THINK OF TUBERCULOSIS

Not only in patients with chronic cough, weight loss and night sweats but also in the following clinical conditions:

• *Acute lobar pneumonia*, especially if the patient does not respond quickly to antibiotic treatment

• *Pyrexia of unknown origin (PUO)* — with or without hepatosplenomegaly or lymphadenopathy

• *Progressive or bizarre neurological syndromes* (consider TB meningitis)

• *Intracerebral space-occupying lesion* (tuberculoma is not as rare as used to be taught)

• *Paraplegia*, whether or not there is a visible gibbus, and even if plain spinal X-rays are normal. (Diffuse spinal arachnoiditis is a cause of paraplegia and may be tuberculous)

• *Chronic diarrhoea* with or without abdominal pain; this may result from tuberculous ileitis

• *Ascites*: a fluid sample must be studied in all cases — if exudate (protein >3.0 g/dl), TB is a possibility and therapeutic trial of antituberculous drugs is usually warranted

• *Normochromic anaemia* with no evident cause

• *Painful indurated skin nodules*, usually in the legs with arthralgia (erythema nodosum)

FURTHER READING

Crofton, J., Horne, N., Miller, F. (1992) *Clinical Tuberculosis*. London: McMillan. (Low-cost edition available in developing countries.)

HIV infection and disease in the tropics

Over 90% of new HIV infections now occur in developing countries. Wherever HIV is prevalent, hospitals and clinics are faced with a large and escalating burden of disease and in several African cities it is already the leading cause of adult death.

HIV has now become a major tropical disease.

Specific guidelines, current seroprevalence data, epidemiological reviews, summaries, reports and updates are published regularly by the Global Programme on AIDS (GPA) which is part of WHO. These are available (usually free of charge) from the GPA Document Centre, WHO, 1211 Geneva 27, Switzerland.

Most countries have national AIDS programmes which provide help and guidance for local and regional projects.

THE VIRUSES

There are two distinct types, HIV-1 and HIV-2, as well as several closely allied species. The simian (monkey) immunodeficiency viruses (SIV), which are endemic in several species of Old World monkeys, are useful experimental models, especially in evaluating possible vaccines. The origins of HIV (both types 1 and 2) are obscure and unlikely ever to be established.

Differences between HIV-1 and HIV-2

There are clear differences between HIV-1 and HIV-2 in genomic structure and in the antibody response to infection. Although not always easy, the two infections can usually be separated serologically. HIV-1 is rapidly spreading round the world and is universally distributed. HIV-2 is much less common and largely restricted to West Africa.

The two viruses are transmitted in the same way but HIV-2 seems less transmissible. Where HIV-1 and HIV-2 coexist, HIV-1 infection is rapidly overtaking HIV-2 in prevalence. Dual infections can occur and there is no evidence that one infection protects against the other.

Both viruses cause the same immune defects and are associated with a similar disease. HIV-2 takes several years longer than HIV-1 to cause significant immunosuppression or death. The main importance of HIV-2 in areas where it is prevalent is to ensure that the kits used for blood tests can detect both viruses. Differentiation is unimportant for individual patient management because treatment is not altered.

The rest of the chapter will refer to HIV without differentiating between the two types.

TESTING FOR HIV

Whom to test?

It is important to have a local policy on HIV testing which reflects local needs, resources and conditions.

As a minimum all blood donors should be tested each time blood is donated, and the hospital should always have sufficient kits to provide this minimum service.

Consent should normally be obtained, and counselling offered where possible. The amount of time that can be devoted to each individual will vary greatly. In areas of high seroprevalence and high demand, counselling may have to be restricted to a single session of less than half an hour.

Patients should only be offered testing if the result is likely to alter management or provide important prognostic information. It may not be necessary to confirm HIV infection in a patient with terminal AIDS. It is unnecessary to test all TB patients unless different drug regimens are used, for example avoiding thiacetazone. Confidentiality is difficult to safeguard when HIV-negative patients receive one therapy and HIV-positive patients another.

Clinical AIDS surveillance alone will provide a falsely low estimate of the impact of HIV disease unless supplemented by HIV testing of blood. All monitoring can be done on stored samples and so can be undertaken without disruption to services.

What test to use?

Some are specific for HIV-1 or HIV-2 whereas others can identify both types. All are highly specific and sensitive if the manufacturer's guidelines are followed correctly; the kits are as accurate as the laboratories using them.

The most widely used tests identify specific anti-HIV antibodies. This can be done by the ELISA method, a variety of rapid and/or simple colorimetric or agglutination tests which do not require a laboratory or electricity, and the Western blot.

There is no single test that is suitable for all circumstances.

ELISA testing is best suited for regular processing of large numbers of samples so that complete plates (usually 90 samples) can be run. It is unsuitable for laboratories with limited facilities and fridge space, irregular electricity supply and inadequately trained and supervised technical staff. With bulk purchasing, WHO can supply ELISA kits that cost £0.40 per test (1995 prices).

Rapid and simple tests such as particle agglutination or dot immunoassay tests can be done in less sophisticated laboratories. Many are available as single kits and so can be efficiently used when small numbers of samples need testing. Some can be stored safely at room temperature. With central purchasing, unit costs can be kept between £0.40 and £2.50, depending on the kit.

Western blots are expensive (over £20), can be difficult to interpret and standardize and are merely serology tests elongated on a piece of blotting paper. The WHO now no longer recommend Western blotting for confirmation, suggesting instead that if it is necessary then the combination of ELISA with a simple or rapid assay is as reliable and much cheaper.

GPA issues regular updates on HIV test kits for use in developing countries. *Operational Characteristics of Commercially Available Assays to Detect Antibodies to HIV-1 and/or HIV-2 in Human Sera; Report 6* was produced in 1993 and is available free of charge from GPA.

EPIDEMIOLOGY

Surveillance

Surveillance is carried out in order to monitor the extent of HIV infection and disease in a given region or community.

In poor tropical countries few resources are available for epidemiological monitoring. At the start of the HIV epidemic in Africa surveillance was only able to show gross changes and to monitor relatively crudely the arrival and subsequent spread of infection. With experience and institutional strengthening surveillance is now much more accurate.

Specific at-risk groups include female sex-workers (prostitutes), attenders at sexually transmitted disease (STD) clinics, workers such as migrant labourers and long-distance lorry drivers, and intravenous drug users. Groups more representative of the general population include pregnant women attending for antenatal care, blood donors, military recruits and newborn infants. It is important to record age and sex in all surveys.

Disease surveillance is usually carried out in hospital and often concentrates on counting cases of AIDS as defined by the WHO in the provisional clinical case definition for Africa. HIV seroprevalence can be measured in specific groups such as hospital admissions, adults with active TB or pneumonia or cadavers in the hospital or district mortuary.

The most valuable data come from surveys that are regularly carried out on the same populations or patient groups and that use the same techniques. In this way trends in seroprevalence and changes in pattern of transmission or clinical disease can be identified.

Seroprevalence

With rapid spread of infection (and delays in reporting and analysis) current figures quickly become out of date.

At the end of 1993 WHO estimated that worldwide more than 14 million adults had already been infected with HIV. Over 9 million (60%) were from sub-Saharan Africa, 1.5 million from Latin America and Caribbean regions and 2 million in south and South-east Asia. At least 2 million of these adults, infected early on in the epidemic, have already died from HIV disease. Some feel that the figures are conservative estimates.

In most areas seroprevalence is still rising. Exceptionally high rates of infection in the general adult population (usually represented by pregnant women in antenatal clinics) have been reached in some urban populations where transmission has been active for several years: 33.9% in Kigali, Rwanda in 1992; 23% in Blantyre, Malawi in 1990; 28% in Kampala, Uganda in 1989–90 and 25% in Lusaka, Zambia in 1990. Seroprevalence in rural areas is lower: 8.2% in south-west Uganda in 1990; 5% in rural Ivory Coast in 1989 and up to 10% in the Kagera region of Tanzania.

In a few areas seroprevalence may already be levelling out and remaining relatively steady. In Kinshasa, Zaïre, seroprevalence appears to have reached about 8% in 1986 and remained relatively constant for at least 3 years.

TRANSMISSION

Sexual transmission

In nearly all tropical countries the most important way HIV is transmitted is by heterosexual sex.

Recently a rate of 0.03 or a 3% chance of acquiring HIV infection after one sexual episode has been described in young men from northern Thailand. The risk elsewhere is unknown.

Risk factors

There are several factors that markedly increase the risk of transmission, the most important being other STDs which cause ulceration, chancroid (*Haemophilus ducreyi*) in particular, as well as primary syphilis and genital herpes simplex. STDs which cause inflammation and discharge, such as gonorrhoea, *Chlamydia* and perhaps trichomoniasis, also increase the chance of sexual transmission of HIV.

For a man, being uncircumcised may be a factor that increases the risk; it certainly increases the risk of acquiring other STDs. Cervical erosion may be a risk factor in women.

Vertical transmission

Of infants delivered from HIV-infected mothers in tropical countries, 25–40% may become infected. Perhaps 30–50% of the viral transmission occurs transplacentally *in utero*; a small fraction become infected during the birth process. The remainder are probably infected postnatally by breast-feeding. The much lower rates of vertical transmission described in industrialized countries (about 10–15% overall) can be explained by the widespread use of formula feeding.

There are no plans to change the universal advice to mothers in developing countries to breast-feed because of the risk of HIV transmission.

Transmission by infected blood

Two groups are at particular risk: patients receiving blood transfusions and people who inject drugs and share needles and syringes. Improperly sterilized injection equipment in hospitals and other health facilities is another (unquantifiable) risk.

Screening blood donors for HIV has greatly reduced the chance of HIV transmission through transfusion. Errors can occur in HIV testing and some donors may be in the 'window' phase with an acute infection which is not yet serologically recognizable.

There is a small risk to health-care workers exposed to HIV-infected blood. In a typical needlestick injury, where the skin is punctured but the inoculum is small, the risk of acquiring HIV is less than 1%.

CONTROL STRATEGIES

A mixture of preventive measures

Several quite distinct ways to intervene seem inherently likely to be effective, although only condom use has been proved to work. They include:

1 Condom promotion and distribution.
2 Effective diagnosis and treatment of STDs.
3 Mass media information and education programmes.
4 Targeted interventions directed to core groups.
5 Campaigns focused on schoolchildren.
6 Individual or peer-group promotion of safe sex.
7 Provision of a safe (screened) blood supply.
8 Needle exchange programmes for drug users.

Condom distribution and STD treatment are the main elements of most comprehensive programmes. Blood safety is a small but important component.

Cost and likely impact

The WHO GPA estimated in 1992 that to implement effective prevention programmes worldwide would cost annually about £1500 million. This is 20 times what was actually spent in 1993 in developing countries but only about 2% of what was spent overall on health care.

The maximum predicted impact of such a strategy is to reduce new infections by about 50%, so, of the projected 19.5 million new cases of HIV expected between 1992 and 2000, 10 million could be prevented, at a cost of £1000 per case avoided. Given the huge direct and indirect costs of HIV disease, prevention may save as much as £50 billion and would be highly cost-beneficial.

Even taking the most optimistic view, it seems certain that HIV disease will remain an important tropical problem for the foreseeable future.

MECHANISMS OF DISEASE

Pathogenesis of HIV infection

The main cell population that HIV infects are lymphocytes that carry the

CD4 antigen on their surface. This is because the CD4 molecule is acting as the receptor to which the virus can initially attach before entering the cell. CD4 lymphocytes are T-helper cells. The key concept in understanding the pathogenesis of HIV infection is the selective loss of function and progressive depletion of T-helper lymphocytes.

T-helper lymphocytes have an important role in the regulation of the cell-mediated immune response and also cooperate with B cells in the production of antibody.

Loss of CD4 cells by HIV infection disrupts both cell-mediated and humoral immunity. This is shown by the loss of delayed hypersensitivity to such skin test (recall) antigens as PPD or tuberculin, *Candida* and mumps antigen; and by polyclonal B-cell activation with hypergammaglobulinaemia.

Other cell populations can be infected, including macrophages which may be important reservoirs of HIV outside the blood and may carry HIV to different organs, including the CNS. Cytokine secretion by infected macrophages is aberrant and may play a role in chronic fever, wasting and enteropathy. Active replication of HIV is evident in lymph nodes at all stages of infection and B cells may be non-specifically activated.

Different strains of HIV may differ in virulence in the cell types that are preferentially infected. A single HIV infection can generate many different antigenic variants; it is possible that in the course of infection variants will eventually emerge that can escape an increasingly exhausted immune system.

Progressive immunosuppression

With the progressive destruction of one part of the immune system a distinct form of immunosuppression develops. As with other immune deficiency syndromes, a relatively limited number of organisms are able to exploit the specific immune defect, and commonly cause disease in seropositive individuals.

It is unclear why some pathogens are so characteristic of HIV and others not. Different pathogens characteristically occur in the early and later stages of HIV disease.

A few conventional pathogens cause clinical disease in the early as well as the later stages of HIV disease: *Mycobacterium tuberculosis*, *Streptococcus pneumoniae*, non-typhi salmonellae (NTS) and the varicella-zoster virus are the most important. Disease is caused by acute infection or by reactivation of a dormant focus.

In the early stages of HIV when immune function is relatively preserved, the main abnormality is a much higher attack rate with clinically typical disease presentation and a normal response to standard treatment. In later stages of HIV and in AIDS itself clinical presentations become atypical and there is a diminished response to standard therapy.

Opportunistic pathogens (relatively avirulent organisms that only usually cause disease in individuals with disrupted immune systems) are only seen in the later stages of HIV disease when much more severe immunosuppression has developed. Immune surveillance is also abnormal and specific cancers can develop.

The opportunistic infections that dominate clinical practice in industrialized countries are *Pneumocystis* pneumonia, disseminated *M. avium*, cytomegalovirus retinitis and toxoplasmosis. The malignancies are non-Hodgkin's lymphoma, primary CNS lymphoma and Kaposi's sarcoma.

These unusual infections and malignancies are so characteristic of (late) HIV disease that they can be grouped together and used to define AIDS.

STAGING HIV DISEASE

CD4 lymphocytes are found in peripheral blood and the normal count in a seronegative person is about 1000 CD4 cells/mm^3. The absolute CD4 count is a useful way of staging HIV infection and assessing the immune status of a patient. But as CD4 counts are seldom available in the tropics, they will not be discussed further.

WHO have proposed a clinical staging system for HIV disease. The four categories are:

Stage 1: Asymptomatic or persistent generalized lymphadenopathy (PGL).

Stage 2: Mild disease with minor weight loss, minor mucocutaneous manifestations, one episode of herpes zoster or recurrent upper respiratory tract infection.

Stage 3: Intermediate disease with weight loss of more than 10% of body weight, unexplained chronic diarrhoea or fever, oral *Candida*, leukoplakia, pulmonary tuberculosis or severe bacterial infection.

Stage 4: Severe disease equivalent to AIDS.

This system may be of some use for individual prognosis but is of little help in clinical management.

NATURAL HISTORY

The following account of illness caused by HIV infection will highlight those areas where the illness in the tropics differs from that reported from north America and Europe.

Acute seroconversion illness

Initially in acute infection most adults develop a high viraemia and a marked fall in CD4 count; they are highly infectious. During seroconversion neutralizing antibodies appear and viraemia is greatly reduced. A minority of adults experience an acute seroconversion illness which resembles glandular fever. A wide range of skin rashes is seen and with transient CD4 depletion oral *Candida* and even opportunistic infections can occur. Because of the relatively non-specific features the seroconversion illness is seldom recognized in tropical countries.

The latent phase

After seroconversion the CD4 count rapidly rises to near normal and the individual feels well. Free virus is difficult to detect in peripheral blood, although active viral replication is taking place in the reticuloendothelial system. Lymph-node architecture is progressively disrupted and nearly 50% of adults develop PGL; this has no prognostic significance.

With time the CD4 count falls and immunosuppression slowly but inevitably progresses. The rate of progression is extremely variable.

It is during this latent infection that most onward transmission of HIV takes place. Latency can be a very difficult concept to convey in counselling.

Early HIV disease

In the relatively early stage of HIV infection, when CD4 counts are only moderately reduced, many individuals start to experience specific symptoms of HIV or will develop a disease typical of early HIV infection.

The HIV symptoms may be weight loss, night sweats, pruritic skin rash, unexplained fever or chronic diarrhoea. Early HIV disease can be relatively trivial, such as oral *Candida* or oral hairy leukoplakia; painful and disabling but not life-threatening, such as herpes zoster; or life-threatening bacterial or mycobacterial infections. This stage of disease is sometimes referred to as AIDS-related complex or ARC.

The serious early HIV diseases are pneumococcal infection (pneumonia and sinusitis), TB (pulmonary and lymphatic) and NTS infections (often bacteraemic). In general, clinical presentation is straightforward and the response to therapy good.

In most tropical regions early disease predominates, whereas it is of relatively minor importance in industrialized countries. This is because of the much higher exposure in poor, overcrowded tropical communities to respiratory and diarrhoeal pathogens.

Late HIV disease or AIDS

The important AIDS-defining opportunistic infections in tropical countries are cryptosporidiosis and isosporiasis in the bowel and cryptococcosis and toxoplasmosis in the CNS.

Extrapulmonary and disseminated TB, severe bacteraemic pneumococcal disease and disseminated salmonellosis are all common. Other Gram-negative septicaemia, especially *Escherichia coli*, are increasingly being recognized. Mixed infections are also frequent.

Early HIV disease is more common than late disease or AIDS in many tropical communities. Much early HIV disease, such as lobar pneumonia or pulmonary TB, is clinically typical and will only be recognized as associated with HIV by serological testing. Studies that just focus on 'clinical AIDS' will fail to identify nearly all early HIV disease.

There are two reasons why early HIV disease predominates. There is intense exposure to TB, the pneumococcus and the salmonellae and few patients are surviving long enough to develop marked immunosuppression and late disease or AIDS. High mortality occurs in the early stages of HIV disease from clinical problems that are easily treated and cured in centres with better facilities and resource. The time from seroconversion to death—usually 10–12 years in developed countries—may be as short as 4–5 years in the tropics.

CLINICAL PROBLEMS

This section will describe the common and important clinical presentations of HIV disease in the tropics and discuss simple investigations, management and treatment. It is assumed that therapy will be limited to a range of cheap broad-spectrum antimicrobials, and that antibiotic sensitivity testing is not routinely available.

Skin problems

Skin problems are common in HIV-positive patients and cause considerable discomfort and morbidity.

GENERALIZED PRURITUS

Many adults develop a chronic itchy maculopapular rash. In dark skins many lesions become hyperpigmented and even nodular.

Some cases are due to scabies, when treatment with benzyl benzoate or lindane will be effective; it may need to be repeated and should include family members. In those who do not respond, antihistamines such as chlorpheniramine are of limited use and may restore normal sleep. A topical corticosteroid sometimes eases itching.

HERPES ZOSTER

Shingles is extremely common and is often the first recognizable HIV-related problem. Where HIV disease is widespread, lay people often recognize the implications of a zoster eruption and realize they are probably HIV-infected. Zoster affecting more than one dermatome is virtually diagnostic of HIV.

Adequate analgesia is important: codeine phosphate 30–60 mg every 6 h is usually effective. Daily dressing may be necessary. Acyclovir is too expensive for poor countries.

SKIN INFECTIONS

Minor wounds can develop into deep-seated and necrotic lesions which may cause bacteraemia. Poor wound healing or unusual skin infections may indicate underlying HIV infection.

If antimicrobials are necessary, the best choice is a penicillinase-resistant penicillin such as flucloxacillin. Erythromycin or chloramphenicol may also be used.

Acute cough and fever

Acute respiratory infections are amongst the most important clinical problems that occur in HIV-infected adults in the tropics. Most patients who present with acute cough and fever will have community-acquired bacterial pneumonia. This has always been an important disease in the tropics but HIV has dramatically increased the incidence of bacterial pneumonia. *Pneumocystis carinii* pneumonia is uncommon.

There is usually a short history of cough, sputum (purulent, rusty or blood-stained), pleuritic pain, fever and rigors, marked toxicity and shortness of breath. Many cases will have had a previous episode.

Examination will reveal a sick and toxic patient, often lying on the side of the pneumonia. The patient is sweaty, taking rapid shallow breaths and coughing frequently. Some patients may be mildly jaundiced, have meningeal irritation and be shocked. Most will have lobar consolidation with bronchial breathing.

Nearly all infections are pneumococcal. *Haemophilus influenzae* is sometimes isolated but in my series from Nairobi, this was usually coinfecting a primary pneumococcal pneumonia. Mixed bacterial/mycobacterial infections are relatively common in HIV disease.

Pneumonia can be confirmed by a chest X-ray but this is wasteful if physical signs are definite. Gram-straining of sputum is useful. Pus cells confirm that a good specimen has been obtained; diplococci should be abundant.

Treatment is the same as for HIV-negative patients: benzylpenicillin 0.6–1.2 g parenterally 6 hourly for 5–7 days. An early switch to oral therapy is kind. If the patient is cyanosed, oxygen (if available) should be given by face mask. If shocked, intravenous fluids are vital.

The prognosis in HIV-related pneumonia is usually excellent. When patients present late with extensive disease mortality is significantly higher.

If the patient is not responding there are several possibilities. There may be coinfection with NTS, *Haemophilus* or TB; an empyema is forming (especially with inadequate initial therapy); or the initial diagnosis is wrong and the patient has pulmonary TB. Changing to chloramphenicol or starting antituberculous therapy may be indicated. As *Pneumocystis* is uncommon in most areas and many salmonellae are resistant, co-trimoxazole is not recommended.

Chronic cough with fever

Chronic respiratory problems are common and highly associated with underlying HIV infection. Most patients have pulmonary TB. The main differential diagnosis will be recurrent or partly treated bacterial pneumonia. Pulmonary Kaposi's sarcoma can occur with skin lesions that are usually obvious. In most tropical areas *P. carinii* is rare.

Chronic cough and fever should be easy symptoms to identify; many patients will have weight loss, night sweats, weakness and haemoptysis; lower-lobe disease is frequent with widespread crepitations; effusions are much more common in HIV. TB frequently recurs so some patients will have had previous therapy.

The diagnosis of TB is more difficult to confirm by radiology or microscopy when there is underlying HIV infection. Classic upper-lobe cavitary disease is much less common and lower-lobe consolidation is more frequent. Fewer cases are smear-positive for acid fast bacilli. TB culture is of little help because the result is so delayed.

Prior to HIV, patients could be left untreated with smear-negative disease because they were a negligible public health risk for onward transmission and would usually return a few months later with smear-positive disease. This cannot be done with HIV-TB and many patients will need to be started on clinical suspicion alone.

National treatment guidelines should be followed. Patients with TB/HIV-related disease respond well to short or standard-course therapy, unless they have end-stage overwhelming disseminated TB.

If bacterial pneumonia is suspected, a therapeutic trial of benzyl-penicillin or ampicillin should be started. If a good response is seen then TB can be excluded. If not, TB therapy should then be initiated.

Thiacetazone toxicity frequently occurs in HIV-positive patients. Some 10–20% may develop skin reactions, usually erythema multiforme, and some progress to Stevens–Johnson syndrome. Because of this, national treatment guidelines have been changed in several countries and thiacetazone has been withdrawn, usually replaced by ethambutol. TB cases can be offered screening and those who are HIV-positive can be given ethambutol.

If there is a limited supply of ethambutol, it can be kept for patients who have started to react. The drug reaction starts with return of fever, itchy skin, non-specific rash then target lesions, usually within 6 weeks of commencing treatment. Stopping thiacetazone at this point will usually prevent severe Stevens–Johnson syndrome from emerging. The main challenge in therapy is to ensure good compliance. Most HIV cases have relatively few organisms in the sputum and are therefore only slightly infectious. A secondary epidemic of TB from the new HIV-associated cases is therefore unlikely.

High fever without focus

Fever is a frequent symptom in HIV-infected individuals. Whilst some patients seem to have no obvious cause for the fever, fever usually indicates infection.

Careful history and examination may sometimes reveal a focus, especially in the CNS, joints or soft tissue; or a cardiac murmur in an intravenous drug-user suggests endocarditis. Patients may have chronic symptoms such as diarrhoea, dry cough or skin lesions and then acutely develop a high swinging fever but no additional focus.

Malaria must always be excluded, although there is no reported interaction between malaria and HIV immunosuppression. Usually a high fever without focus in the tropics in a patient with underlying HIV indicates a high-grade bacterial or mycobacterial infection.

In HIV-infected patients this clinical presentation is sometimes referred to as an enteric fever-like illness and is very common. NTS rather than Salmonella typhi are important. Disseminated TB is increasingly recognized but Mycobacterium avium is rare.

Without blood culture, salmonella bacteraemia cannot be reliably diagnosed. In Africa about 10% of all HIV-positive adults presenting to hospital will have NTS bacteraemia and a further 5% may have other Gram-negative sepsis (including E. coli and classical S. typhi). Disseminated TB will only be diagnosed postmortem.

The next problem is how to differentiate the two, and this cannot be done clinically. A positive blood culture will obviously identify Gram-negative sepsis but a negative culture may occur if antibiotic therapy has recently been given, or it may indicate disseminated TB. Many cases of disseminated TB will be anergic with minimal pulmonary lesions. *Mycobacterium tuberculosis* can take weeks to be positive in a blood culture, unlike *M. avium*.

How can these sick patients be managed without any microbiology support? My experience is that, with awareness of the possibility of NTS bacteraemia, prompt broad-spectrum antibiotic therapy backed up by intravenous fluids can reduce an 80% mortality to about 25%. Knowing the antimicrobial sensitivity pattern can further reduce mortality to about 15%.

In marked contrast, I have known very few HIV-positive patients with *M. tuberculosis* mycobacteraemia and none with *M. avium* in Africa to survive, even with the prompt use of quadruple therapy. Disseminated mycobacterial disease seems to be an agonal problem in Africa, and may be so in other tropical regions.

This suggests the following strategy. For any patient with an enteric fever-like illness, exclude a silent pulmonary TB focus by a chest X-ray. Then concentrate on the treatment of NTS bacteraemia. First-line blind therapy can be ampicillin and gentamicin, or chloramphenicol. A quinolone would be the best choice but may not be available or affordable.

If there is little or no improvement after 3 or 4 days (like typhoid itself, NTS bacteraemia can take several days to respond), drug resistance may be a problem. Switch to whichever first-line treatment was not used initially.

Chronic diarrhoea

'Slim disease' is what many people equate with African AIDS. It was the first clinical problem specifically associated with HIV infection by Ugandan investigators in 1985, and named by the local patients and their carers, who recognized a major new disease in their community.

Chronic diarrhoea with profound wasting is easy to identify and is highly associated with underlying HIV infection. Widespread metastatic disease, advanced TB (consumption), Addison's disease and untreated insulin-dependent diabetes present with wasting but not usually diarrhoea. Whilst diarrhoea and wasting are the most obvious problems on the wards in Africa, if comprehensive studies of hospital admissions are done they account for only 10–20% of the HIV-related workload.

In some hospitals the number of 'slim' patients is dropping as carers recognize that hospitals have little to offer. Home care is increasingly seen as a more appropriate option.

The diarrhoea is usually painless, watery and without blood or mucus. It can be variable or intermittent. Profound weight loss is clinically obvious and may be 20% or more of the premorbid weight. High fever is not typical and should suggest NTS septicaemia or agonal disseminated TB.

Studies from various African centres have shown a variety of stool pathogens: about 15% will have *Cryptosporidium*, 15% enteropathic bacteria, 10% *Isospora* and 10% widespread TB with faecal mycobacteria. Stool viruses may be found in some but microsporidium is probably rare. Amoebae, *Giardia* and helminths are not important. No pathogen is identified in many cases. The hyperinfection syndrome with *Strongyloides stercoralis* is not associated with HIV infection.

Specific treatment is very limited. Some salmonellae and *Shigella* and *Isospora* will respond to high-dose co-trimoxazole, so a trial of therapy may be indicated. There is no effective treatment for cryptosporidium and disseminated TB is unlikely to respond. Codeine phosphate 30 mg three times daily may offer some symptomatic relief.

Many patients have painful oral *Candida*. Antifungal agents are all expensive, but gentian violet mouthwashes may give some relief.

Prognosis in patients with 'slim' disease is variable but usually only of the order of several months. Profound weight loss at presentation suggests death in weeks only.

It is now almost universally acknowledged that 'slim' disease is best managed by a home-care programme.

Other specific problems

KAPOSI'S SARCOMA

Only 4% of the first 5000 reported AIDS cases in Uganda had Kaposi's sarcoma. Although the cancer is endemic in east Africa, the high rates noted in homosexual men in the USA have not been seen. Kaposi's sarcoma may be caused by an unidentified sexually transmissible agent that is more readily passed by anal rather than vaginal intercourse.

It is usually easy to diagnose clinically. Unmistakable purple or violet raised plaques and sometimes nodular lesions can occur anywhere in the skin. The roof of the mouth is a typical site. The lesions are painless and do not itch. Lymphadenopathy is common and lesions can disseminate, particularly to the lungs. Biopsy is not necessary.

Some patients with few lesions progress slowly and will die from other causes. Other patients progress rapidly over several months. Widespread pulmonary disease is an ominous sign. The drugs used for chemotherapy are expensive, and palliative only.

ENCEPHALITIS

The presentation may suggest a generalized encephalitis. A relatively mild headache and an altered mental state are most common. Other signs include hemiparesis, ataxia, cranial nerve lesions, generalized incoordination, seizures and confusion. Fever is variable.

The most likely cause in the tropics is cerebral toxoplasmosis. One autopsy study from west Africa found evidence of cerebral toxoplasmosis in 15% of HIV-infected cadavers and was considered a prime cause of death in 10%. However, in Kenya experience is that clinically obvious encephalitis is rare (less than 5%).

The cost of specific treatment (£30/week) rules out its use in developing countries.

Paediatric disease

Little work has been done on paediatric HIV disease in the tropics. It seems likely that there will be a strong association with pathogenic bacteria, particularly *Streptococcus pneumoniae*, *Haemophilus influenzae*, NTS and *M. tuberculosis*: they are already recognized as important problems in children with HIV infection in industrialized countries.

One would expect higher rates of otitis media, primary TB, acute lower respiratory disease, acute and chronic diarrhoeal disease and perhaps pyogenic meningitis.

Many children with late HIV disease present with failure to thrive and malnutrition. Wasting rather than oedema is seen and the child will often have an associated maculopapular or nodular rash with active excoriation (consider scabies) and generalized lymphadenopathy. The HIV-positive child will not respond to nutritional rehabilitation.

Many centres with limited resources now promote early discharge of such children because the prognosis is so poor. The mother will need counselling, even if not HIV-tested.

PROPHYLAXIS AND ANTIRETROVIRAL THERAPY

Antiviral therapy

AZT is far too expensive to use widely in the tropics and seems likely to remain so.

TB chemoprophylaxis

The use of a 6-month course of INH as TB prophylaxis in HIV infection is being studied. No conclusions have yet been reached.

Preventing pneumococcal disease

It is premature to recommend pneumococcal vaccination for anyone with HIV infection in the tropics until efficacy data are available and sensible recommendations can be made.

Other infections

There are no NTS vaccines licensed for human use. There seems no point in offering *Pneumocystis* prophylaxis in populations who only rarely develop the problem. Similar arguments can be made for *M. avium* infection and *Toxoplasma* prophylaxis.

FURTHER READING

Ansary, M.A., Hira, S.K., Bayley, A.C., Chintn, C., Nyaywa, S.L. (1989) *A Colour Atlas of AIDS in the Tropics*. Wolfe Medical Publications, London.

De Cock, K. *et al.* (1990) AIDS — the leading cause of adult death in the west African city of Abidjan, Ivory Coast. *Science* **249**, 793–6.

De Cock, K. *et al.* (1992) Tuberculosis and HIV infection in sub-Saharan Africa. *JAMA* **268**, 1581–7.

Gilks, C.F. (1993) The clinical challenge of the HIV epidemic in the developing world. *Lancet* **342**, 1037–9.

Gilks, C.F. *et al.* (1990) Life-threatening bacteraemia in HIV-1 seropositive adults admitted to hospital in Nairobi. *Lancet* **336**, 545–9.

Gilks, C.F. *et al.* (1992) The presentation and outcome of HIV-related disease in Nairobi. *Q J Med* **267**, 25–32.

Janoff, E.N. *et al.* (1992) Pneumococcal disease during HIV infection: epidemiologic, clinical and immunological perspectives. *Ann Intern Med* **117**, 314–24.

Lucas, S.B. *et al.* (1993) The mortality and pathology of HIV infection in a west African city. *AIDS* **7**, 1569–79.

Merson, M.H. (1993) Slowing the spread of HIV: agenda for the 1990s. *Science* **260**, 1266–8.

Potts, M. *et al.* (1991) Slowing the spread of HIV in developing countries. *Lancet* **338**, 608–23.

Tamashiro, H. *et al.* (1993) Reducing the cost of HIV antibody testing. *Lancet* **342**, 87–90.

Fevers in general

Once the obvious causes of fever in an individual patient have been eliminated, it may be very difficult to find the cause of the fever. This difficulty is even greater in areas where laboratory facilities are very limited. The daily fever pattern is only helpful in certain cases, and cannot be made the basis of a clinically useful classification.

ACUTE OR CHRONIC FEVER?

A start can be made by calling fevers of more than 2 weeks' duration chronic fevers. There are far fewer cases of chronic fever than there are of acute fever. Simple laboratory tests combined with experience will enable most fevers to be identified, and a correct decision taken about treatment. Even the rather imprecise diagnosis of viral fever is worth something to the patient. When uncomplicated it predicts a favourable outcome and that there is no need for antibiotic treatment. The WBC is often very helpful in identifying the likely cause of a fever.

Routine investigations in all fevers

1 Thick blood film for malarial parasites.
2 Total and differential WBC.

In holoendemic malarial areas, positive blood films are so common it is often very difficult to decide whether the malaria parasites are related to the fever or not. If in doubt, treat the malaria and see what happens. When malaria is causing the fever, the WBC is usually normal or reduced and the platelet count is often low.

If the WBC shows a polymorphonuclear leucocytosis (PNL), even if the blood film is positive for malaria, it is likely that there is another cause for the fever.

Malaria is the most important cause of fever in the world. It may cause acute, chronic or relapsing patterns.

ACUTE FEVERS

Acute fevers with a negative malarial blood film

The WBC divides this group into two, as is shown here:

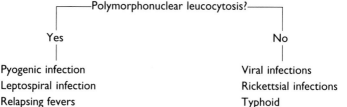

Yes	No
Pyogenic infection	Viral infections
Leptospiral infection	Rickettsial infections
Relapsing fevers	Typhoid
Amoebic liver abscess	
Acute collagenosis	

Acute fevers with PNL and localizing symptoms

Organ-specific symptoms or signs often localize the site of infection, e.g.:
1 Severe sore throat:
   (a) streptococcal tonsillitis;
   (b) diphtheria.
2 Cough, pleuritic pain, rusty sputum: pneumonia.
3 Severe pain and swelling in a joint: pyogenic arthritis.
4 Frequency, dysuria and loin pain: pyelonephritis.
5 Severe pain in the head and back of the neck with neck stiffness: meningitis.
6 Severe pain in a bone: osteomyelitis.
7 Severe lower abdominal pain: pelvic sepsis.
8 Bloody diarrhoea: bacillary dysentery, *Campylobacter* infection.
9 Marked localized lymphadenopathy: local sepsis, plague.
10 Sharply defined cutaneous inflammation: erysipelas.
11 Ill-defined subcutaneous inflammation: cellulitis.
   The delay in seeking medical aid so common in tropical countries greatly increases the chances of the patient presenting with a useful localizing sign or symptom.

Acute fevers with PNL and no obvious localizing features

The most important are:
1 Septicaemias of all kinds: most commonly staphylococcal, streptococcal and meningococcal, including acute bacterial endocarditis.

2 Infections with *Leptospira* and *Borrelia* spp.:
    (a) leptospirosis (may be jaundice, renal involvement);
    (b) tick-borne relapsing fever;
    (c) louse-borne relapsing fever.
3 Acute non-typhoid *Salmonella* septicaemia. Gastrointestinal symptoms may be few. Rose spots and splenomegaly may develop.

Acute fevers without PNL with a negative blood film

The most important are viral and rickettsial infections and typhoid fever. Most viral fevers are never diagnosed precisely, because of their relatively non-specific features, unless the physician is helped by a good virology laboratory. Features suggesting a viral aetiology are:
1 The double-humped fever pattern described under dengue (see p. 78).
2 Many other cases seen in the same space/time envelope.
    In some of the rickettsial infections, localized adenopathy may draw attention to a previously unrecognized eschar. In typhoid, persistent complaints of abdominal discomfort or pain draw attention to an abdominal location of infection.

Important non-localizing accompaniments of fever

These include:
1 Rash.
2 Generalized or multiple adenopathy.
3 Splenomegaly.
4 Anaemia.
5 Jaundice.
6 Polyarthritis.
    Some patients have localizing features and non-localizing features in addition to fever. These features are just as important in long-term as in short-term fevers. Sometimes the diagnosis unfolds itself as time passes. The following is an example.

Acute fever with a tender liver

I was called to see a British engineer in a nearby hospital who had returned from a tour of duty in Asia a week previously. He gave a 4-day history of fever with severe pain over the liver. There was no nausea or vomiting, the temperature was 39°C and the liver was enlarged to 2 cm below the costal margin and extremely tender. He was not jaundiced. What diagnoses do you think most likely?
    The extreme tenderness of the liver, in the absence of jaundice, and

a past history of several attacks of bloody diarrhoea in the last few months, made me suspect amoebic liver abscess (ALA), possibly very superficial, so that a small abscess (or abscesses) would cause the pain early. But he could have had viral hepatitis. What investigations would you request?

1 WBC and differential, malarial blood film. (The total count was 6400/µl, with a relative monocytosis; blood film negative.)

2 Liver function tests (awaited).

3 An ultrasound scan of the liver. (This led to suspicions of early abscess formation on the anterior surface of the right lobe.)

4 Amoebic sensitized particle agglutination test (SPAG). (This was negative – it often is in the early stages of ALA but has the advantage of being simple and available commercially.)

5 ESR (10 mm/h).

Are these results conclusive? No, but the WBC is very atypical of ALA, as is the normal ESR.

Two days later the patient became deeply jaundiced and the liver function tests were typical of viral hepatitis. His serum was positive for hepatitis B surface antigen and anticore IgM, indicating acute hepatitis B. The severe pain and tenderness rapidly resolved and he made an uneventful recovery.

In tropical areas the hepatitis B surface antigen carrier rate is at least 40 times higher than in west Europe. Partial immunosuppression produced by malaria may be a factor. Hepatitis B is a major cause of chronic active hepatitis, cirrhosis and hepatoma in the tropics. There is now great interest in vaccination strategies to prevent both vertical and horizontal transmission of the infection, especially now that an effective bioengineered vaccine is available (Engerix B).

Acute fever with a haemorrhagic rash

There are two main causes:

1 Viral haemorrhagic fevers (dengue, Chikungunya, Rift Valley fever).

2 Acute meningococcal septicaemia.

Correct diagnosis is vital, because death may follow within a few hours of petechiae appearing in meningococcal infection. Material aspirated from the spots contains meningococci in more than 80% of cases when examined by a simple stained smear. Other pyogenic septicaemias cause purpura only rarely.

Non-haemorrhagic rashes accompany a great number of acute fevers. One of the few general infections in which a rash never occurs is malaria, and even then a rash may be precipitated by a drug given to treat the attack. I have seen a gross ecchymotic rash in a Japanese patient with falciparum malaria in Thailand treated with an injection of amidopyrine.

Acute fever with anaemia

The main causes are:
1 Malaria.
2 Babesiosis.
3 Bartonellosis.
4 Infection in a patient with a pre-existing anaemia such as sickle-cell disease or thalassaemia, or with G6PD deficiency.

Causes (1) and (4) are much the commonest. Polyarthralgia is usual in sickling crises, which are often precipitated by infections. Jaundice often accompanies pneumonia in patients with G6PD deficiency.

PROLONGED FEVER

By this we mean fever of more than 2 weeks' duration. There is some overlap with the acute fevers, because sometimes infection with the same organism causes fever for a week, sometimes for a few weeks. Typhoid fever in some individuals causes a mild fever of only a few days' duration, in others a continuous fever for 3 weeks. Also, infection with *Borrelia* may cause only one bout of fever or several.

The following list contains the commonest causes of prolonged fever, simply subdivided according to the most usual WBC picture:
1 Chronic fever with PNL:
   (a) deep sepsis
   (b) ALA   } typically a sustained fever;
   (c) erythema nodosum leprosum
   (d) cholangitis   } typically a relapsing pattern.
   (e) relapsing fever
2 Chronic fever with eosinophilia:
   (a) invasive (toxaemic) *Schistosoma mansoni* and *japonicum* infections;
   (b) invasive *Fasciola hepatica* infection;
   (c) acute lymphangitic exacerbations of *Wuchereria bancrofti* and *Brugia malayi* infections;
   (d) gross visceral larva migrans due to *Toxocara canis*.
3 Chronic fever with neutropenia:
   (a) malaria;
   (b) disseminated tuberculosis;
   (c) visceral leishmaniasis;
   (d) brucellosis.

4 Chronic fever with normal WBC:
  (a) localized tuberculosis;
  (b) brucellosis;
  (c) secondary syphilis;
  (d) trypanosomiasis;
  (e) toxoplasmosis;
  (f) subacute bacterial endocarditis (SBE);
  (g) systemic lupus erythematosus (SLE);
  (h) chronic meningococcal septicaemia.
5 Chronic fever with a variable WBC picture:
  (a) tumours;
  (b) reticuloses;
  (c) drug reactions.

I have missed out those conditions where the localizing signs are so obvious that they could not really be overlooked, such as pyogenic arthritis.

Chronic fevers with a relapsing pattern

In these, a period of fever is typically followed by an afebrile period, after which the fever recurs. Some of the above infections may show this feature but the following very often do:
1 Malaria.
2 Visceral leishmaniasis.
3 Trypanosomiasis.
4 Relapsing fever.
5 Brucellosis.
6 Filariasis.
7 Cholangitis.

Pyrexia of undetermined origin in the tropics

This is a difficult problem at any time, and is even more difficult in the tropics because of the usual lack of sophisticated techniques for investigation. One often has to rely on very simple investigations combined with clinical acumen. Even in the UK, the problem may be very difficult, as the following case illustrates.

A young, female Indian doctor working in the UK presented with a 6-week history of fever, anorexia and weight loss. She was referred by a chest physician who had investigated her thoroughly with entirely negative findings. Among the negative results were:
1 Full blood count, including repeated films for malaria.
2 Chest X-ray.

3 Urine examination.
4 Repeated blood cultures.
5 Liver biopsy, including cultures for TB.
6 Marrow, including cultures and animal inoculation.
7 Intravenous pyelography.
8 Mantoux test to 1, 10 and 100 tuberculin units (TU).
9 *Brucella* cultures and antibody titres.
10 *Toxoplasma* dye test.
11 Rickettsial and viral antibodies.
12 Blood for lupus erythematosus (LE) cells and antinuclear factor.
The only significant findings were:
1 ESR 80 mm/h.
2 Slight lymphocytosis in the differential WBC.

Two weeks later she was continuing to run a fever of 37–39°C, and had lost a further 3 kg in weight. She had no specific symptoms. Her temperature rose every evening, and she spent long periods weeping because she feared she was going to die. What would you do next?

(Her mother had been suffering from chronic pulmonary TB for 15 years, and this had been treated with each new antituberculous drug as soon as it was released. It seemed likely that the patient had been exposed to drug-resistant bacilli.)

Eight weeks after the onset of her illness, she started treatment with rifampicin, ethambutol and INH in conventional doses. Three days later her temperature was normal for the first time since she fell ill. After a further 2 weeks' treatment she had regained her appetite and much of her lost weight. On the day she was sent back to the referring physician for continuation of her treatment at home, her chest X-ray showed typical miliary mottling. Her Mantoux became strongly positive* a few weeks later.

The commonest cause of PUO in the tropics is cryptogenic TB, and by this I mean disseminated TB without evidence of its origin. The presence of a negative Mantoux supports the diagnosis, as in many tropical areas it is unusual to find an adult who is not Mantoux-positive, and disseminated disease is accompanied by a generalized suppression of cell-mediated immunity. Once the overwhelming infection is overcome, the test becomes positive, often very strongly.

Prolonged fever with a normal WBC and ESR

There are two main causes of this—one is spurious (the patient is simulating fever) and the other follows.

---

* It ulcerated with 10 TU.

A previously fit Yorkshire businessman presented with a 2-week history of fever ever since returning from a successful business trip to Brazil. His only complaints were of flu-like symptoms and myalgia.

Physical examination was negative apart from a pulse rate of 110 beats/min and temperature of 38.4°C. All the usual investigations were initiated. Two weeks later all the results received were negative. In particular, the blood films were negative for parasites, the WBC was normal, the ESR was normal, and the chest X-ray and urine results were normal. But he continued to have fever with tachycardia, and muscle pains had increased. What parasite could cause this picture?

The answer is *Toxoplasma gondii* infection (toxoplasmosis). He had eaten raw, minced steak (steak tartare) in Brazil. His *Toxoplasma* dye test was positive at a titre higher than the local public health laboratory had ever recorded. He responded rapidly to treatment with a sulphonamide combined with pyrimethamine and did not relapse.

T. gondii can be acquired from eating the oocysts excreted by cats harbouring the sexual stages of the parasite. But outbreaks have been traced to the inadequately cooked infected meat of herbivores: in his case it was raw beef. Undercooked beef or mutton has been incriminated in several previous outbreaks.

Prolonged fever and HIV infection

HIV infection must come into the differential diagnosis nowadays, because the virus itself can cause fever, and opportunistic infections—especially due to salmonellae, pneumococci and TB—may also be responsible.

Prolonged fever with ascites

A 7-year-old Bedu girl presented in a Gulf state with a history of fever of many weeks' duration and swelling of the abdomen. On examination she was febrile (temperature 38.6°C) and the abdomen was prominent with signs of free fluid and rather vague intra-abdominal masses. What is the most likely diagnosis?

The answer is tuberculous peritonitis. This was confirmed by finding tubercle bacilli in the peritoneal fluid, and her prompt response to specific treatment. Can any other forms of intra-abdominal TB give rise to chronic fever?

Yes, the following:

1 Ileocaecal TB.

2 Tuberculous abdominal glands.

3 Tuberculous enteritis.
4 Renal TB.

TROPICAL FEVERS: CONCLUSIONS

Most prolonged fevers in the tropics can be diagnosed by fairly simple
tests. When all tests fail, one may have to resort to therapeutic trial.
There is nothing to be ashamed of about this, provided one remembers
its limitations.

Diseases commonly presenting as diarrhoea

CHAPTER 12

Amoebiasis

Amoebiasis is an infection, usually of the colon, caused by *Entamoeba histolytica*. It is found in all parts of the world where environmental sanitation is poor. The organism may behave as a parasite (by harming the host) or as a commensal (when it does no harm to the host). It is a disease of human beings — although some monkeys may have it and the monkey infection is transmissible to humans.

PARASITE AND LIFE CYCLE

Humans swallows the four-nucleated cyst in food or water contaminated by human faeces. The cyst is digested in the gut to release several small amoebae. These live in the colon, normally on the surface of the mucosa. They feed on bacteria and other food residues. Usually they cause no harm; so of every 5 people swallowing amoebic cysts, only 1 develops amoebic dysentery, but 3 others will develop asymptomatic infection. This has been confirmed by experiments in volunteers.

The amoebae multiply in the gut by simple binary fission. As the amoebae move around the colon from right to left, the colonic contents become more solid. In the more solid contents, the actively motile amoebae stop feeding, empty their food vacuoles and become rounded. They then secrete a cyst wall that makes them resistant to drying up. These amoebic cysts passed in the formed stool of people *without* amoebic dysentery are the infective agents.

All cysts begin life as amoebae. Infection is by swallowing cysts. Cysts are never found in the tissues. If gastrointestinal transit time is shortened from any cause, amoebae will be found in the stools of a patient previously passing cysts. So finding amoebae in the stools of a patient with diarrhoea does not prove that the diarrhoea was *caused* by the amoebae. (Koch originally observed *E. histolytica* amoebae in the stools of patients with cholera.) The amoebic trophozoites are said to be non-infective but not when inoculated directly into the colon. An epidemic of amoebic dysentery was caused by an incorrectly functioning enema machine used by chiropractors in the USA, resulting in several deaths.

The adult amoeba (trophozoite) is variable in size, highly motile by means of its pseudopodia and characteristic flowing motion, and when in an aggressive stage, it usually contains ingested red cells.

The cysts are spherical, and when mature contain four typical nuclei and sometimes a glycogen mass and a refractile chromidial bar. Only when cysts are more than 10 μm (and less than 18 μm) in diameter are they called *E. histolytica*. Smaller cysts may lead to asymptomatic infection with smaller, non-pathogenic amoebae called *E. hartmanni*. Not all strains of *E. histolytica* are capable of tissue invasion. In highly specialized laboratories, pathogenic strains can be distinguished from non-pathogenic strains with a high degree of confidence by studying their isoenzyme patterns on electrophoresis.

PATHOGENESIS

Amoebic dysentery

From time to time the amoebae, previously browsing harmlessly on the surface of the colonic mucosa and feeding on bacteria and debris, may become invasive. When they do this they breach the mucosa with the aid of cytolytic enzymes, change their diet by consuming the erythrocytes escaping from damaged capillaries and cause ulcers.

When infection is with a pathogenic strain, the amoeba can change its behaviour in the same host from time to time, due to factors which are little understood. There is no doubt that immunosuppression of the host can turn a harmless commensal infection into a dangerously invasive one. It seems possible that other causes of bowel damage, such as an intercurrent bacterial infection or *Trichuris* or *Schistosoma mansoni* infection, may have the same effect.

The ulcers in the colonic mucosa, usually described as flask-shaped because they tend to undermine the mucosa and so have a narrow neck on cross-section, are responsible for the symptoms of amoebic dysentery. Amoebae in the tissues do not, by themselves, excite an inflammatory response; nor do they release toxins causing a constitutional illness.

TYPICAL CLINICAL PICTURE

Amoebic dysentery

The diarrhoea is not usually of sudden onset; the incubation period can seldom be determined, but is probably not less than 7 days and commonly weeks or months.

The diarrhoeal stool often contains macroscopic blood and mucus, and its frequency is seldom more than 12 per day; the appearance is obviously faecal. It is not usual for the patient to have severe abdominal pain or tenesmus, but low abdominal cramps often precede defecation. There is usually no significant fever or malaise. The main differences between amoebic and bacillary dysentery are outlined in Table 12.1.

NATURAL HISTORY OF UNTREATED AMOEBIC DYSENTERY

In the absence of treatment, the usual course is for the diarrhoea to become gradually worse over a period of a few days, persist for a few weeks and then gradually diminish in intensity so that the whole bout is over in about 6 weeks.

Attacks may recur at irregular intervals for years; in the intervening intervals the infection persists in commensal form. This type of infection exactly mimics the course of non-specific ulcerative colitis. Sometimes there are no attacks of dysentery after the first, but the persistence of the infection manifests itself years later as a complication. On the other

AMOEBIC AND BACILLARY DYSENTERY

Bacillary dysentery	Amoebic dysentery
May be epidemic	Seldom epidemic
Acute onset	Gradual onset
Prodromal fever and malaise common	No prodromal features
Vomiting common	No vomiting
Patient prostrated	Patient usually ambulant
Diarrhoea watery, bloody	Diarrhoea faecal, bloody
Stool odourless	Stool: fishy odour
Stool microscopy: few bacilli; pus cells; macrophages; red cells	Stool microscopy: numerous bacilli; red cells; amoebic trophozoites with ingested red cells
Abdominal cramps very common, sometimes severe	Occasional low abdominal cramps
Tenesmus common	Tenesmus uncommon
Natural history: spontaneous recovery in a few days to a week or more; no relapses	Natural history: bout lasts several weeks; dysentery recurs after a variable interval of remission; infection persists for several years

Table 12.1 Differences between amoebic and bacillary dysentery.

hand, some patients have persistent bloody diarrhoea of unremitting intensity which lasts for years.

Most of those infected, even if untreated, probably eliminate the infection in 3–5 years, unless they are reinfected in the meantime.

INVASIVE VERSUS NON-INVASIVE DISEASE

If a patient is excreting E. *histolytica* cysts or trophozoites, he or she has amoebiasis. If there are relevant bowel symptoms he or she obviously has invasive disease. On the other hand, if there are no symptoms he or she may have either non-invasive disease or subclinical invasive disease. This last situation arises when the patient has ulcers so few or so small that they cause no symptoms, but the tissues are nevertheless being invaded by the amoebae. This concept has some relevance to treatment.

LOCAL COMPLICATIONS

Occasionally, severe bleeding may occur when a large blood vessel is eroded or perforation may lead to peritonitis. Both these events are rare. So also is fulminating amoebiasis, in which the amoebae are so destructive that the colon is destroyed and the patient presents with severe abdominal pain and ileus. This picture is that of a surgical abdominal emergency and at laparotomy the colon may be found to be necrotic, when survival is unusual. Severe but less fulminating cases also occur, and the distinction from bacterial colitis is not always easy.

Amoebiasis and vague abdominal symptoms

Where amoebiasis is common (40% of the population may be infected), there is a tendency to blame any abdominal complaint on the amoebae. Unless invasive disease can be demonstrated, there is little scientific support for this view, but if specific antiamoebic therapy is followed by recovery, it seems reasonable to accept an amoebic aetiology.

Strategic amoebic ulcer

Ulceration in or near the rectum may cause tenesmus and the over-secretion of mucus. The condition may be persistent.

Amoeboma

Amoeboma is the name given to a chronic inflammatory mass developing around a part of the bowel, caused by E. *histolytica*. It seems highly likely that secondary infection plays an important role, because amoebae by themselves do not excite an inflammatory response. The mass often develops as a diffuse thickening of the bowel wall, closely resembling a carcinoma. Its mechanism is not properly understood, but it may originate in a subacute perforation. It may affect any part of the large bowel – most commonly the ileocaecal region. The symptoms are commonly those of subacute obstruction, namely abdominal pain and constipation, sometimes with fever. The diagnosis is usually made by the pathologist, following surgical removal. If the diagnosis is made before surgery, the response to specific antiamoebic treatment is usually rapid.

Postdysenteric colitis

The authors have personal experience of almost 1000 patients, ex-Far-East prisoners of war of the Japanese, who suffered repeated attacks of amoebic dysentery 30 or so years before review. They mostly had no specific treatment for the 3.5 years they spent in captivity. In no single case were there strictures or other serious sequelae of the infection, although a considerable proportion complained of persistent looseness of the stools in the intervening years. It seems from this that serious damage to the bowel is an unusual sequel of amoebic infection, even when it is allowed to progress for long periods without treatment. Nevertheless, that it does sometimes occur is well documented.

Amoebic liver abscess

PATHOLOGICAL PROCESS

During an attack of amoebic dysentery or years after repatriation from an area endemic for amoebiasis, even in the absence of any history of dysentery, or at any time in a person living in such an area, an amoebic liver abscess may develop. It is caused by amoebae entering the liver via the portal vein, and probably usually develops from the coalescence of several small abscesses. Most abscesses develop in the right lobe. An area of liver-cell destruction occupies the centre, containing an amorphous, cream, pink or pinkish-brown viscous fluid called amoebic pus – or anchovysauce pus, by those who enjoy comparing pathological fluids with foods. Pus which is at first pale on aspiration often darkens on exposure to air. Surrounding the pus there is an area of compressed but otherwise normal liver tissue in which the amoebae, as trophozoites, are feasting on the liver cells and actively dividing. Beyond this zone the liver is normal, and no inflammatory cells or capsule contain the abscess. The abscess, unlike most others, contains not degenerate pus cells, but the products of colliquative necrosis of the liver cells.

If not interfered with, the abscess will normally grow until it reaches a surface through which it can discharge, such as the skin, the periton-eum, the pleural cavity or the pericardium. The stretching of the liver capsule produced by the abscess is presumably the main source of pain; the absorption of necrotic tissue causes the characteristic constitutional disturbances.

COMMON SYMPTOMS
1 Pain over the liver (sometimes referred to the right shoulder).
2 Fever.
3 Tenderness over the liver noticed by the patient.
4 Cough if there is associated lung involvement.
5 Breathlessness in chronic cases with marked anaemia.

PHYSICAL SIGNS
1 Enlargement of the liver.
2 Tenderness over the liver.
3 Localized oedema or tenseness of the skin when the abscess approaches the surface.
4 Signs at the right lung base: dullness to percussion and reduced breath sounds, perhaps with crackles.

COMPLICATIONS
1 Rupture, leading to:
  (a) formation of a cutaneous sinus;
  (b) amoebic empyema;
  (c) hepatobronchial fistula (and secondary pulmonary lesions) or death from inhalation of abscess contents;
  (d) amoebic peritonitis;
  (e) amoebic pericarditis (perhaps with eventual constrictive pericarditis).
2 Metastasis with secondary lesions in lung or brain.
3 Jaundice, although marked jaundice is rare.
4 Anaemia, which may be severe if the abscess is very chronic.

Cutaneous amoebiasis

This spreading, destructive condition, involving all layers of the skin and often the deeper tissues, may develop after prolonged contact with amoebae from any cause. It is particularly common when an amoebic liver abscess drains through the skin, as a result of spontaneous rupture or surgical drainage, if no specific antiamoebic treatment is given. Gross perianal and perineal destruction may follow uncontrolled amoebic dysentery, when the skin is bathed in a culture of virulent amoebae.

Gruesomely destructive amoebic infections of the penis may result from anal intercourse with a partner with amoebiasis, and comparable lesions may affect the vulva.

DIAGNOSIS

Amoebic dysentery

DIRECT DIAGNOSIS

The amoebae are sought by microscopic examination of a saline smear, using a $\times10$ objective. A freshly passed stool specimen must be examined because, on cooling, the amoebae stop moving (their movement is their most easily recognizable characteristic) and empty their food vacuoles which, containing red cells, provide another useful diagnostic feature. The 'rounded-up' amoebae present in stale stool specimens are very difficult to identify. The amoebae are usually most abundant in bloody or mucoid portions of the specimen.

In frank amoebic dysentery the amoebae can usually be found easily. In less severe cases of invasive intestinal amoebiasis, the amoebae are most easily found in curettings taken from ulcers at endoscopy.

Amoebae can be grown in culture, but the method is of little clinical use.

Amoebae in biopsy material do not stain well with haematoxylin and eosin, and a special stain to reveal nuclear detail such as iron–haematoxylin is often necessary for positive identification.

INDIRECT DIAGNOSIS

Up to 90% of cases of amoebic dysentery have demonstrable specific antibodies in their serum. Several methods are used, the IFAT being the commonest. This method can also be used to identify E. histolytica in tissue sections. Not only do frozen sections give good results but haematoxylin and eosin-stained sections, treated with fluorescein-conjugated antiamoebic serum, may show specific immunofluorescence years after the original section was cut. The unpredictability of this method is probably explained by variations in the fixation methods employed.

CIRCUMSTANTIAL DIAGNOSIS

There are no very helpful findings based on general investigations. In all cases of suspected amoebic dysentery, efforts should be focused on finding the parasites.

Intestinal amoebiasis in the absence of dysentery

DIRECT DIAGNOSIS

If the stools are formed, the search is for cysts, not for amoebae. To exclude the diagnosis with a high degree of certainty, several stool specimens have to be examined, by direct microscopy and by a sensitive concentration technique. We accept three negative specimens as adequate; more conservative physicians insist on 10 or more. The figure arrived at depends partly on one's confidence in the laboratory. This is because cyst passage is irregular, and the numbers passed vary greatly from day to day. To avoid incorrect identification, it is essential for the cysts to be measured accurately, such as by an eyepiece micrometer, for the cyst of the non-pathogenic amoeba *Escherichia coli* may, if it contains four nuclei or fewer, closely resemble the cyst of *Entamoeba histolytica* in all features other than size. Do not believe laboratory reports which do not specify cyst size. Any laboratory which cannot measure cyst diameter is inevitably unreliable.

When cysts are seen in unstained preparations, their morphological features should be confirmed in wet preparations using Lugol's iodine or in dried and fixed smears using iron–haematoxylin.

It is not helpful to purge the patient before examining the stool.

INDIRECT DIAGNOSIS

Only 40% of cases with asymptomatic amoebiasis have demonstrable antibodies, rendering this test virtually useless in such cases, bearing in mind its inability to distinguish between present and past infections.

The low seropositivity rate is probably because many such infections involve non-pathogenic strains which do not invade the tissues.

Why bother searching for cysts?

The search for *E. histolytica* cysts is mainly a European and North American preoccupation. In countries where amoebiasis is not endemic, there is a natural concern to ensure that patients, even in the absence of symptoms, are not harbouring an infection which could lead to serious complications later. In endemic areas, it is questionable if it is worth treating asymptomatic infections. This is because the overall prevalence rate may approach 40%, with the chances of rapid reinfection correspondingly high. It is worth giving a luminal amoebicide to someone who has had an amoebic liver abscess, but this is done whether or not cysts are found in the stools.

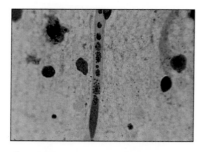

Plate 1

Plate 2

Plate 3

Plate 4

Plate 5

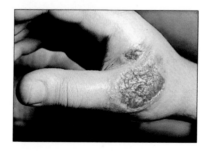

Plate 6

Plate 1 Impression of brain (Giemsa stain) showing a capillary 'blocked' by schizonts in fatal cerebral malaria due to *Plasmodium falciparum*.

Plate 2 Ring haemorrhages in fatal cerebral malaria: presumably these lesions are irreversible.

Plate 3 A 'leishmanioma' developing after accidental innoculation of *Leishmania donovani* in the laboratory. There was no dissemination of infection following the local excision of the lesion.

Plate 4 *Leishmania tropica* lesion (Saudi Arabia).

Plate 5 Cutaneous leishmaniasis lesion due to L. tropica (Afghanistan).

Plate 6 Cutaneous leishmaniasis lesion due to L. mexicana (Belize).

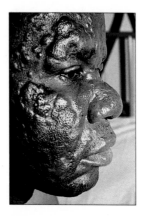

Plate 7

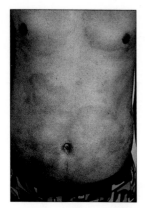

Plate 8

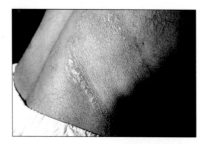

Plate 9

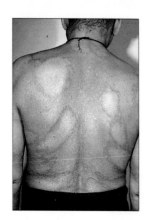

Plate 10

Plate 7 Lepromatous leprosy—nodular form (Uganda).

Plate 8 Borderline lepromatous (BL) leprosy (Bangladesh). This patient relapsed with BL disease after dapsone monotherapy of lepromatous leprosy. Dapsone resistance was demonstrated in the bacilli responsible for relapse.

Plate 9 Tuberculoid leprosy (India). The single lesion is anaesthetic, scaly and dry, and has a raised edge.

Plate 10 Borderline tuberculoid leprosy. There were numerous anaesthetic lesions, and extensive nerve damage. The patient eventually went blind due to the neurological complications.

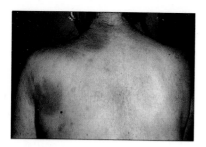

Plate 11

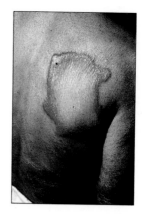

Plate 12

Plate 13

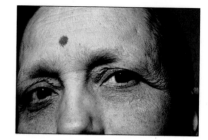

Plate 14

Plate 11 Erythema nodosum leprosum (Indian patient from Uganda). The patient presented with numerous painful red swellings, fever and a high leucocytosis. She had been treated for cellulitis with penicillin before the diagnosis was finally made. She required 60 mg prednisolone a day to control the symptoms, and the disease process took more than a year before it subsided.

Plate 12 Upgrading (reversal) reaction. The previously flat lesions suddenly became hot, painful and raised above the surface of the surrounding skin.

Plate 13 Upgrading (reversal) reaction. The nose became completely obstructed. The reaction developed within 4 weeks of initiating chemotherapy, and rapidly resolved on steroid treatment (prednisolone 40 mg daily) without permanent sequelae. (Severe and irreversible nerve damage may occur in such cases.)

Plate 14 Madarosis in lepromatous leprosy. There were no localized lesions in the skin whatsoever, but the skin of the entire body seemed slightly thickened, with an exaggerated pore pattern approaching 'peau d'orange' (orange-skin appearance). Skin-slit smears—as nasal smears—were all strongly positive for *Mycobacterium leprae*.

Plate 15

Plate 16

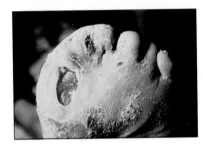

Plate 17

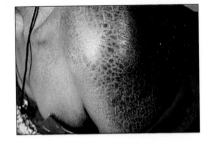

Plate 18

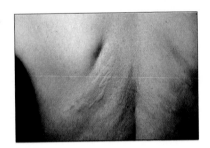

Plate 19

Plate 15 Visibly thickened posterior auricular nerve. The patient had a 'tuberculoid'-type lesion on the palm of his right hand, and an associated thickening of the dorsal branch of the radial nerve. As in this case, not all leprosy patients obey the rules of 'classification'.

Plate 16 Combined ulnar and median nerve palsies in a patient with lepromatous leprosy. Unfortunately his hobby is karate.

Plate 17 Perforating plantar ulcer in a patient with lepromatous leprosy and extensive damage to the major nerve trunks. She lost her other leg following a perforating plantar ulcer which led eventually to amputation. Nerve damage can progress even when the *Mycobacterium leprae* is under satisfactory control from effective chemotherapy.

Plate 18 Clofazimine skin changes. This is the skin of a patient (India/East Africa) whose dose was increased to 300 mg/day when she developed erythema nodosum leprosum (ENL). However, when the dose was reduced to 100 mg/day the skin changes persisted, and her conjunctivae also developed pigmentation.

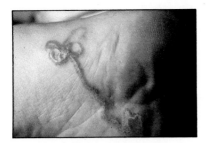

Plate 20

Plate 22

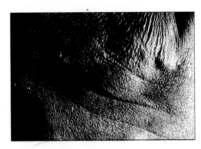

Plate 21

Plate 23

Plate 19 (*Opposite*) The typical rash of *Strongyloides stercoralis* seen in fair-skinned people. The serpiginous wheals come and go in a few hours, hence 'larva currens'. There is no other skin rash like it. (British prisoner of war infected in Thailand.)

Plate 20 In contrast, this is larva migrans due to the dog hookworm *Ancytostoma brasiliensis*. The lesions last for many weeks, and the active edge advances only a few millimetres a day (Barbados).

Plate 21 Early *Onchocerca* dermatitis. The patient was a Nigerian nurse. DRB treated her in Nigeria with diethylcarbamazine (DEC), and she relapsed whilst on a course in Liverpool. This was not really surprising as DEC has no activity against the adult worms, which continued to produce microfilariae in the meantime.

Plate 22 Early onchocerciasis with an itchy papular rash in a student from Cameroun.

Plate 23 Onchocerciasis. A 40-year-old Nigerian farmer with nodules over the lateral chest wall, an itchy dermatosis, and 'presbydermia'; an appearance of ageing of the skin caused by destruction of elastic tissue from the liberation of toxins by dying microfilariae.

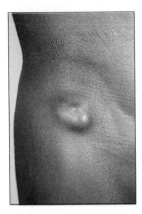

Plate 24

Plate 25

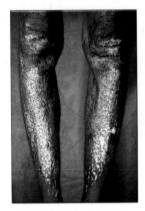

Plate 26

Plate 27

Plate 24 An *Onchocerca* nodule. The typical site in Africa. Many worms may be incarcerated in a multiloculated nodule.

Plate 25 Advanced onchocerciasis (West Nigeria). There is depigmentation of the skin overlying the shins, enlargement of the inguinal lymph glands associated with laxity of the surrounding skin (hanging groins) and generalized presbydermia.

Plate 26 Onchocerciasis: typical depigmentation associated with grossly exaggerated skin fold pattern.

Plate 27 Mazzotti reaction. This woman's skin looked entirely normal before she was given 100 mg diethylcarbamazine. She complained of fever and of severe pruritus.

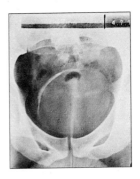

Plate 28

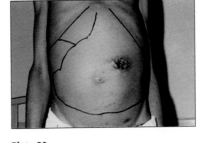

Plate 29

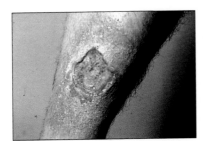

Plate 30

Plate 31

Plate 28 Calcified bladder in *Schistosoma haematobium* infection. Many such bladders are capable of entirely normal function, the calcification involving the eggs rather than the bladder tissues.

Plate 29 Visceral leishmaniasis in a Kenyan with massive splenomegaly. The burn to the abdominal wall was from traditional treatment.

Plate 30 An acute tropical ulcer in a European (République de Guinée).

Plate 31 A chronic tropical ulcer in a 16-year-old Nigerian. Fibrosis has raised the ulcer above the level of the surrounding skin. Treatment at the local dispensary had been incompetent (repeated acriflavine gauze dressings) and caused so much pain that the patient had defaulted.

There is a strong case for identifying and treating asymptomatic amoebiasis, even in areas of high endemicity, during pregnancy and when immunosuppressive therapy is planned. The eradication of asymptomatic infection in these instances could well prevent severe invasive disease.

It is essential to exclude amoebiasis in patients who have ever visited an endemic area before embarking on corticosteroid treatment of ulcerative colitis, or any other condition where corticosteroids are indicated.

Amoebic liver abscess

DIRECT DIAGNOSIS

As the lesion is closed, the only access to diagnostic material is by aspiration. Amoebae are sought in the last portions of the aspirate, and will be found in about 80% of cases, if the search is enthusiastic and sustained.

But diagnostic aspiration is no longer justified, now that a non-toxic drug is available for treatment. Aspiration is nowadays used for therapy, not for diagnosis. For this reason, the diagnosis of ALA usually depends on indirect or circumstantial evidence.

INDIRECT DIAGNOSIS

Antibodies can be detected in almost all cases of ALA (95–100%). Many methods are used, including indirect fluorescence, immuno-electrophoresis and sensitized particle agglutination. The level of anti-body in ALA is usually high, provided the condition has been established for more than a week or two. This helps to distinguish a positive result in ALA from the positive results found in a high proportion of residents in an endemic area who have low levels of antibodies, presumably resulting from previous intestinal infections.

A commercial test kit (the latex particle agglutination test marketed by Ames, under the name Serameba) gives results that correlate well with high titre results of more sophisticated tests, such as the IFAT. It is too expensive for use in most developing countries, and has a short shelf-life. But the test is so simple to perform and contains such good instructions that it requires no special technical expertise. Control positive and negative sera are provided with each kit.

Antigen-detection tests can be used with considerable accuracy to identify the presence of amoebae in stools or tissues, but these have not yet been developed to the point where they are helpful clinically.

CIRCUMSTANTIAL DIAGNOSIS

In most parts of the world, this method of confirming a clinical diagnosis is the most important, because material for direct diagnosis is usually inaccessible and serodiagnostic methods are usually unavailable. The typical positive, but non-specific findings, are:

1 Leucocytosis (a polymorphonuclear leucocytosis occurs in more than 80% of well-established abscesses).

2 Anaemia. This again is a feature of established abscesses. It is usually normochromic and normocytic but in very chronic abscesses hypochromia may develop due to failure of the marrow to utilize iron.

3 A high ESR. A normal ESR almost excludes ALA. The ESR is usually above 50 mm/h, more often around 100.

4 A raised hemidiaphragm and other radiological signs. The diaphragm is usually raised in the right lobe abscesses, and the diaphragmatic movement on respiration reduced. This is best revealed by asking the patient to sniff during screening. There may be an effusion at the right base, consolidation or atelectasis at the right base, or apparently unconnected shadowing in the lung.

5 In liver function tests about 10% of patients have a raised conjugated bilirubin, in keeping with obstructive jaundice. Other liver function tests seldom show consistent or significant changes.

6 Isotope or ultrasonic liver scans will reveal one or more areas of discontinuity in the liver substance. Their relative sensitivity varies with the details of the techniques used, but neither technique is likely to reveal discrete lesions less than 0.5 cm in diameter.

7 Stool examination. In an endemic area so many people are excreting cysts that finding cysts in a patient suspected of having an abscess is of little significance. In any area of the world, the failure to find cysts is of no help in excluding ALA, because they are only found in about half of all cases of proven abscess. Of all the circumstantial diagnostic tests for ALA, stool microscopy is the least helpful.

TREATMENT

The aim of drug treatment is to kill all amoebae in the tissues, and eradicate amoebae from the bowel lumen. Amoebicides are usefully classified by their site of action.

Tissue amoebicides

These drugs act systemically, and kill amoebae in close contact with tissue fluids:

1 Metronidazole (orally), the only drug effective in both the tissues and (to a lesser extent) in the lumen of the colon. A formulation (expensive) for intravenous use is also available.

2 Emetine (parenterally).

3 Dehydroemetine (parenterally) appears less toxic than emetine, mainly due to its more rapid excretion.

4 Chloroquine (orally), is only really effective in the liver, where it is selectively concentrated.

All except chloroquine give a symptomatic cure of amoebic dysentery. Only metronidazole gives a fair chance of parasitological cure—but only if used in high doses.

Luminal amoebicides

These drugs act on amoebae in the lumen of the gut. They may act directly on the amoebae, or indirectly by suppressing the bacteria on which the amoebae depend.

1 Direct-acting luminal amoebicides: diloxanide furoate (Furamide); emetine bismuth iodide; various iodoquinoline compounds (e.g. di-iodohydroxy-quinoline, clioquinol).

2 Indirect-acting luminal amoebicides: most important are the incompletely absorbed tetracyclines: tetracycline, oxytetracycline, chlortetracycline. Paromomycin (Humatin (PD)), works but is very expensive and has no role in developing countries.

Treatment of amoebic dysentery

SPECIFIC DRUG TREATMENT

Routine

Metronidazole 800 mg three times a day for 5 days, plus Furamide 0.5 g three times a day for 5 days.

Severe cases

Add tetracycline 500 mg three times a day for 5 days, plus an injection of emetine hydrochloride 65 mg or injection of dehydroemetine 100 mg intramuscularly or subcutaneously daily until the diarrhoea subsides (this is very seldom for more than 5 days).

If you do not have metronidazole for dysentery, use an injection of emetine or dehydroemetine (as above) and a contact amoebicide.

If you do not have Furamide, use one of the iodoquinoline drugs mentioned: a 3-week course is usually needed.

High doses of metronidazole (800 mg three times a day) often cause nausea and vomiting, and many people cannot tolerate this dose. In this case, use 400 mg three times a day for 5 days, and give diloxanide furoate 0.5 g three times a day for 10 days. The lower dose of metronidazole will usually give prompt clinical cure, but the diloxanide furoate is needed to ensure that the infection is eradicated.

In endemic areas, stool examinations for test of cure are largely a waste of time. In non-endemic areas, at least three negative stool examinations, carried out about 4 weeks after treatment is completed, are needed to prove parasitological cure. With combined metro-nidazole—diloxanide furoate treatment as described, this is probably a counsel of perfection.

Treatment of ALA

SPECIFIC DRUG TREATMENT

Routine

Metronidazole 400 mg three times a day for 10 days. This will almost always sterilize the abscess but cannot be guaranteed to eradicate the bowel infection. To do this, give a course of a good contact amoebicide such as Furamide (0.5 g three times a day for 10 days).

Note: Metronidazole is no more effective in ALA than the old combination of emetine and chloroquine, but it is less toxic. If you do not have Furamide, give high-dose metronidazole as for amoebic dysentery.

DRUG FAILURE IN ALA

Failure to respond to metronidazole in the dosage recommended is probably because of high pressure in the abscess preventing the drug reaching the amoebae in the abscess wall via the circulation. It seems that high failure rates are associated with reluctance to aspirate.

There is no doubt that aspiration, by helping the blood-borne drug to penetrate the abscess wall, can enhance the effect of amoebicidal drugs. In Thailand in the Second World War, when emetine was very scarce, many patients recovered from ALA when treated with a single injection of 65 mg emetine hydrochloride, provided the abscess was aspirated, repeatedly if necessary.

INDICATIONS FOR ASPIRATION OF ALA

I Localized swelling or bulging of the ribcage or abdominal wall.

2 Marked local tenderness or oedema.

3 A very raised diaphragm.

4 Failure to respond to conservative treatment.

The aim is to prevent spontaneous rupture. Use a wide-bore needle, syringe and three-way tap. Go through the right chest wall, or the abdomen if the abscess is very superficial. Be guided by scans, X-rays or the point of maximum tenderness. Several aspirations may be needed. There may be more than one abscess. Always aspirate to dryness. A meticulous aseptic technique must be observed.

Treatment of asymptomatic amoebiasis

In tropical areas with a high prevalence of infection, treatment is seldom given for asymptomatic infections and is largely a waste of time. The best treatment is probably Furamide 0.5 g three times a day for 10 days. If Furamide alone fails, the course can be repeated, combined with a course of metronidazole as for dysentery (these 'failed' cases, although asymptomatic, presumably have subclinical invasive disease).

EPIDEMIOLOGY

In endemic areas the infection is usually sporadic rather than epidemic, so much so that if an epidemic is reported the diagnosis should be suspect. Gross faecal contamination of water supplies is needed to produce water-borne spread, because the number of cysts excreted by a cyst-passer is several orders of magnitude smaller than the number of typhoid bacilli produced by a typhoid carrier.

Sporadic infections usually result from faecal contamination of food, such as when human excreta are used as fertilizer for vegetables eaten raw, or by contamination of food by dirty fingers. Again, contamination must be gross because, in contrast to the situation with *Salmonella* spp., *E. histolytica* cannot multiply in the infecting medium.

Amoebiasis is not restricted by latitude: outbreaks have been reported in Indian settlements in Canada and in Oslo in the winter. The main requirements for the transmission of the infection are poor personal hygiene and poor environmental sanitation. Such conditions may be reproduced even in non-endemic areas, in closed institutions harbouring people with a low standard of hygiene, such as the mentally handicapped.

High prevalence rates of *E. histolytica* reported in northern Europe (e.g. Scotland from findings of cysts in stools) have been based on incorrect identification, *E. hartmanni* being mistaken for *E. histolytica*.

Autochthonous cases of symptomatic *E. histolytica* are so rare in the UK that proven cases are usually formally published. Many authorities assume that such patients acquired their infection from a carrier harbouring an exotic pathogenic amoeba. There are no effective measures in the UK or other European countries to ensure that food-handlers who have lived in endemic areas are not infected with *E. histolytica*.

Control of amoebiasis

High standards of personal hygiene (largely a function of health education) and high standards of environmental sanitation (the task of hygienists) are the main enemies of amoebiasis. Personal protection in endemic areas can be achieved by avoiding hazardous foods (such as salads and other cold foods in hostelries) and by drinking only boiled water or bottled drinks. People living in endemic areas can safely eat salad vegetables provided they are washed thoroughly and sterilized in a sodium hypochlorite solution such as Milton, according to the instructions given on the bottle.

Drugs for prophylaxis of amoebiasis

Drug prophylaxis is widely used to prevent diarrhoea in travellers, usually with more benefit to large drug companies than to travellers. In the past the repeated and prolonged use of clioquinol (Enterovioform) for this purpose was eventually shown to be complicated by a serious neurological disorder, subacute myelo-optic neuropathy syndrome, in some individuals.

There is no suitable drug prophylaxis for amoebiasis, and although various antibiotic regimens have been shown to reduce the risk of other (bacterial) causes of travellers' diarrhoea, the consensus is that prophylactic medication should not be a usual recommendation for travellers.

QUESTIONS, PROBLEMS AND CASES

1 You have a patient (a 30-year-old farmer in the tropics) with a 5-week history of fever, pain in the right hypochondrium and an enlarged tender liver.

(a) With this information only, list the possible causes of his illness.

(b) What are the most useful investigations to help you decide the cause of his illness?

2 You have admitted a patient to your hospital with a clinical diagnosis of amoebic dysentery. Would you isolate him? Give the reasons for your answer.

3 What is the differential diagnosis of a tender mass in the right iliac fossa in a 20-year-old female patient in the tropics?

4 A 25-year-old Englishman returned from a 3-week holiday in north Africa 2 days ago. Five days before his return he had high fever, malaise, severe abdominal pain and vomiting associated with the sudden onset of bloody diarrhoea. His general symptoms settled in 4 days, but he still has bloody diarrhoea. Stool microscopy reveals the presence of actively motile trophozoites of *E. histolytica*, containing red cells.

(a) Is a diagnosis of amoebic dysentery adequate to explain his illness?
(b) What other investigation is needed?
(c) What treatment should he have at this stage?

5 A 12-year-old African boy presents with a 2-week history of bloody diarrhoea without fever. On examination his liver is palpable 2 cm below the costal margin and is tender. His stool contains haematophagous trophozoites of *E. histolytica*. He is not jaundiced and liver function tests are normal. The ESR is 55 mm/h and the WBC is normal. What is the most likely cause of the liver findings?

6 A 15-year-old African youth is admitted to a small mission hospital in west Africa with a 6-week history of pain over the right hypochondrium and fever. He has a low fever, and the liver is enlarged, hard and irregular in texture. Investigations show haemoglobin 10 g/dl, WBC 14 000 µ/l, 70% neutrophils, ESR 55 mm/h; the blood film shows scanty *Plasmodium falciparum* parasites; and the stool microscopy shows eggs of hookworm and *Ascaris lumbricoides* present. There is no X-ray machine and no facilities for α-fetoprotein estimation. What would you do?

FURTHER READING

Case records of the Massachusetts General Hospital: Case 19 (1990) *N Engl J Med* **322**, 1378–85.

Chuah, S.-K. *et al.* (1992) The prognostic factors of severe amebic liver abscess: a retrospective study of 125 cases. *Am J Trop Med Hyg* **46**, 398–402.

Editorial (1985) Is that amoeba harmful or not? *Lancet* **1**, 732–4.

Mandal, B.K., Schofield, P.F. (1992) Tropical colonic diseases. *BMJ* **305**, 638–41.

Giardiasis

PARASITE AND LIFE CYCLE

Giardiasis is an infection of the upper small bowel with a flagellate protozoon *Giardia lamblia* (*G. intestinalis*) which may cause diarrhoea. The trophozoite stage of the parasite is a flattened, pear-shaped creature about 15 μm long, 9 μm wide and 3 μm thick. It is concave on its ventral surface where it attaches itself, by its sucking disc, to the intestinal mucosa. It has four pairs of flagella for locomotion. It multiplies in the gut by binary fission. When trophozoites drop off the duodenal and jejunal mucosa and are carried down the gut in the intestinal contents they usually encyst. The cyst is oval, 8–14 μm long by 5–10 μm wide, and contains four small nuclei and a central refractile axostyle. The cysts are infective as soon as passed. When swallowed by a new host, they excyst in the upper gastrointestinal tract and liberate trophozoites. The infection is spread from human to human by the faecal–oral route. Humans are the only important reservoir of infection but wild beavers have been found to be infected in north America. Cysts can survive outside the body for several weeks under favourable conditions. The infection is cosmopolitan, but commonest in parts of the world where standards of environmental sanitation are low.

CLINICAL PICTURE

After swallowing cysts for the first time, symptoms commonly develop 2–6 weeks later. In some patients the infection is asymptomatic, and in others it entirely fails to become established. The factors controlling susceptibility to the infection are poorly understood but very severe infections may develop in immunodeficient hosts. In highly endemic areas, symptomatic infections are seldom seen in adults. In other areas, symptoms may develop in infected subjects of all ages.

The cardinal symptom is diarrhoea. This is usually of subacute onset — but sometimes very acute — and it continues for weeks or months in the absence of treatment. Stool frequency is typically 3–8 times a day; the stool is usually pale, offensive, rather bulky and accompanied by much flatus. There is no blood or mucus, and abdominal pain is restricted to cramps accompanying the urgent call to stool. Tenesmus does not occur, but perianal soreness may develop. Abdominal distension is typical, and patients often complain of bloating, borborygmi and flatus. Burping of offensive or sulphurous gas is a famous but not invariable symptom.

Anorexia is usual in the early stages, and there may be vomiting. Most patients with symptomatic infections lose weight, mainly because of poor appetite, although malabsorption is doubtless sometimes a factor.

PATHOGENESIS

This is not properly understood. There are non-specific changes on jejunal biopsy, mainly in the form of mucosal infiltration with round cells. Some cases show varying degrees of villous atrophy, but this may well be associated with bacterial colonization of the small gut. It has been suggested that malabsorption, often demonstrated by increased excretion of fat and impaired absorption of xylose, is due to the mechanical effects of large numbers of adherent parasites obscuring the surface of the mucosa. The difficulty of getting biopsy material fixed without shedding the trophozoites may account for the contradictory views on this point.

Natural history

Because up to 20% of residents in tropical areas may be excreting *G. lamblia* cysts in their stools and most of them are asymptomatic, and because very few of them have ever had specific treatment, one must assume that the clinical effects of the infection are usually self-limiting. A patient with a normally competent immune system probably loses the symptoms spontaneously in a few months, and thereafter continues to harbour the parasite without any apparent ill-effects. It is unclear whether this is due to an acquired ability to limit the number of parasites or to the development of immunity to a concomitant bacterial infection of the small bowel. Occasionally, even in patients without other signs of immune deficiency, the infection leads to severe inanition in which life is in jeopardy unless something is done. This is particularly a problem in young children with severe malnutrition, in whom there may develop a vicious cycle of impaired absorption due to giardiasis and impaired ability to eliminate *Giardia* because of malnutrition.

DIAGNOSIS

Direct diagnosis

Diagnosis is usually made by finding the characteristic cysts in a direct saline smear preparation of the stool, and the detection rate is improved by using a good concentration technique such as the modified formol-ether method. The number of cysts bears no relationship to the severity of the patient's symptoms, and those with the most severe symptoms sometimes apparently excrete no cysts. This may be because in such cases almost all the trophozoites are adhering to the mucosa. In children with severe diarrhoea, lively, motile trophozoites are quite often found in the stools.

When the diagnosis is strongly suspected on clinical grounds, but no cysts can be found in the stool, trophozoites may be recovered by duodenal aspiration or by using the 'hairy string test' (Enterotest, Hedeco Corporation, USA).

Indirect diagnosis

There is no routine method in use for detecting antibodies to G. lamblia but early experimental serodiagnostic methods are encouraging. Even if perfected, this method will not replace direct diagnosis. Methods of detecting antigen in the stool (e.g. ELISA) have proved sensitive and specific, but are not yet in general use.

Circumstantial diagnosis

No non-specific findings distinguish giardiasis from other conditions causing diarrhoea or malabsorption, but in comparison with coeliac disease and tropical sprue, the development of anaemia is uncommon.

Therapeutic trial

If the diagnosis is strongly suspected on clinical grounds, but no parasites can be found, a therapeutic trial with metronidazole is justified. Because of its activity against anaerobic bacteria, it may cause temporary improvement in tropical sprue, and so yield inconclusive results.

TREATMENT

METRONIDAZOLE

Either 400 mg three times daily for 5 days or 2–2.5 g once daily for 3 days.
These are doses suitable for adults of 50 kg body weight and above. Alcohol may produce flushing, headache, and other unpleasant side-effects in patients taking metronidazole. Patients should be warned to avoid alcohol until treatment is completed. The dosage should be reduced in children accordingly.

TINIDAZOLE

A single dose of 2 g is usually effective.

MEPACRINE

100 mg three times daily for 5–7 days. The dosage should be reduced according to body weight in children.

FURAZOLIDONE

10 mg/kg daily for 5 days in divided doses (maximum of 600 mg/day). It is available as tablets of 100 mg and as suspension. It may cause haemolysis in patients with G6PD deficiency.

ALBENDAZOLE

Cure rates equal to those of metronidazole (95%) have been achieved using 400 mg daily for 5 days.

Treatment failure

If treatment with one drug fails, one can either give a repeat course or change to another drug. A failure rate of about 10% is usual. Sometimes repeated relapses are due to reinfection from another member of the family with asymptomatic infection. In this case, all infected members of the family have to be treated simultaneously.

Spurious treatment failure

If diarrhoea persists after treatment of giardiasis, it is not always due to failure to eliminate the infection. Some patients may have secondary lactose intolerance following giardiasis, and some may have other concomitant disease such as tropical sprue. It is advisable to make sure that the patient does still harbour the infection before giving repeated courses of treatment directed at the parasite.

EPIDEMIOLOGY AND CONTROL

Infection is usually sporadic and in highly endemic areas the ratio of asymptomatic to symptomatic cases is very high. Infection is usually acquired there in childhood and followed by a high degree of tolerance.

Epidemics occur where there is gross contamination by cysts of water supplies used by non-immune populations, such as tourists visiting St Petersburg, or in the crew of a ship taking on contaminated water. Cysts may remain infective in water for up to 3 months.

The situation in some Russian cities might be due to 'anastomoses' between the sewage and water systems produced by heavy bombing and shelling in the Second World War. Several epidemics in the USA have been traced to sewage-contaminated well water. Cysts are not killed by the normal chlorination process.

Control is by environmental sanitation and attention to personal hygiene. In the UK the prevalence is low in the indigenous population but some transmission does occur in this country, especially in residential communities of children and the mentally handicapped. Others at special risk are overland travellers to the Far East.

Personal protection involves the avoidance of raw vegetables and tap water in endemic areas. Boiling destroys the cysts rapidly, and an effective filter removes them.

FURTHER READING

Gilman, R.H. *et al.* (1988) Rapid reinfection by *Giardia lamblia* after treatment in a hyperendemic third world community. *Lancet* 1, 343–5.

Meyer, E.A. (1985) The epidemiology of giardiasis. *Parasitol Today* 1, 101–5.

Meyer, E.A. (ed.) (1990) *Giardiasis. Human Parasitic Diseases*, vol. 3. Elsevier Science.

Balantidiasis

BALANTIDIUM COLI INFECTION

The parasite is a motile, large (50–70 μm long by 40–50 μm wide), ciliated creature which lives in the colon of humans, pigs and rodents. Cysts form when diarrhoea subsides and the rectal contents become formed. The cyst, ingested by a fresh host, excysts to liberate the trophozoite. Multiplication is by binary fission in the trophozoite stage. Human infection is usually from pigs and is rare.

Illness of *B. coli* infection

The trophozoites cause colonic ulceration, just as those of *Entamoeba histolytica*, but perforation is more common. Metastatic lesions do not occur.

DIAGNOSIS

This is by finding the large, majestically motile trophozoites on microscopy of direct-smear saline preparations of freshly passed stool.

TREATMENT

Any of the older (partially absorbed) tetracycline preparations, in a dose of 1–2 g daily in divided doses for 5–10 days.

EPIDEMIOLOGY AND CONTROL

The disease occurs worldwide where humans have close contact with pigs, and has been reported from Ireland as well as from tropical countries. Human to human transmission occurs in special circumstances such as in closed communities. Its control obviously depends on improved standards of environmental cleanliness, and in particular the rigorous exclusion of pig faeces from the human environment.

Cholera

Cholera is a bacterial infection of humans caused by *Vibrio cholerae* (of classical or El Tor biotypes) which characteristically causes severe diarrhoea, and death (in those severely affected) from water and electrolyte depletion. Spread is directly from person to person by the faecal–oral route, or indirectly by infected food or water. It can spread to any part of the world, and may become endemic where standards of environmental sanitation and personal hygiene are low.

Humans are the only reservoir of infection. The El Tor biotype has now largely displaced classical cholera as the major pathogen of public health importance.

CLASSICAL DISEASE AND ITS MECHANISM

After an incubation period, usually in the range of 1–5 days, there is diarrhoea of rapid onset. After the colon is evacuated of faecal material, profuse painless diarrhoea follows. The diarrhoea is typically watery, white and flecked with mucus–the infamous 'ricewater stool'. In 80% of cases, vomiting follows soon after the diarrhoea.

Fever is unusual except in children, and short-lived. The temperature is usually subnormal when the patient is first seen. The speed of onset depends on the size of the infecting inoculum and the natural defences of the host. Gastric acid is a very effective barrier to infection. If it fails, the vibrios multiply very rapidly in the alkaline medium of the small intestine, and a specific toxin called choleragen causes the breakdown of the functional barrier between the intestinal epithelium and the blood. The result is the outpouring of large volumes of protein-free fluid with an electrolyte composition similar to that of plasma, although the potassium (average 13 mmol/l) and bicarbonate (average 44 mmol/l) concentrations are both higher than in plasma. (These values refer to adults. In children the potassium loss is about twice as great, but the sodium loss–about 100 mmol/l–is less.)

There is no significant inflammation of the gut, and the epithelium remains morphologically intact. The illness is self-limiting and diarrhoea ceases in a week if the patient survives. Death is usually due to hypovolaemic shock and acute renal failure.

CHANGES IN FLUID AND ELECTROLYTES

Dehydration is caused by the profuse diarrhoea, compounded by the usual inability to retain fluids by mouth. The speed with which severe dehydration occurs is greater than in any other disease. Collapse from hypovolaemic shock may occur within a few hours of the onset, and the untreated patient may die within 24 h. Hypoglycaemia is common, especially in children.

Apart from profound asthenia accompanying the fall in circulating blood volume and blood pressure, there are obvious signs of fluid depletion in the tissues. These include loss of turgor in the cheeks leading to a pinched appearance, sunken eyes due to orbital dehydration, general loss of skin elasticity detectable by the delay in a pinched skinfold returning to its normal position, and shrivelling of the skin of the fingers – 'washerwoman's hands'. Most people are familiar with this appearance as the result of staying in the bath or swimming pool for too long. In severe cases cerebration is impaired, the voice is weak and husky, and urine output is reduced or ceases altogether.

Cramps of the muscles of the limbs and abdomen are a typical feature of severe cases. They are very painful, and result from a reduction in the concentration of calcium and chloride ions. The loss of water is relatively more than the loss of electrolytes in children, who often suffer hypertonic dehydration as a result.

CLINICAL SPECTRUM

The condition described is the classical picture, seen in a minority of infections only. In an outbreak, for every case of classical disease, there will be at least 10 other cases with mild or asymptomatic infections.

DIAGNOSIS

Clinical diagnosis is only possible in classical cases with profuse, painless diarrhoea, ricewater stools, gross dehydration and muscle cramps. Only occasionally do other infections produce this picture. A certain diagnosis can only be made by isolating the organism.

Direct diagnosis by microscopy

In severe cases the gut is rapidly emptied of its normal contents, and the colon rinsed out by the fluid pouring from the small bowel. In such cases the watery stool contains cholera vibrios in almost pure culture. They can be recognized with a high degree of confidence by dark-field examination of a wet preparation. Identification can be confirmed by adding specific antiserum, which immobilizes the vibrios immediately. Microscopic identification is much more difficult in milder cases because of the presence of large numbers of normal faecal organisms. In asymptomatic cases direct microscopy is useless.

Fluorescence microscopy, using a fluoroscein-conjugated specific anticholera serum, is more sensitive than dark-field microscopy, but more complicated.

Direct diagnosis by culture

Various media are used for primary isolation. Among the best is TCBS (thiosulphate–citrate–bile salt–sucrose) agar. Small numbers of vibrios can only be detected using an enriched liquid medium such as alkaline peptone water. The specimen is best taken by a sterile rubber catheter inserted into the anus, or by a rectal swab. Specimens taken in the field can be transported to the laboratory in sealed plastic bags, or after inoculation into a holding medium.

Indirect diagnosis

After recovery from an attack of cholera, various antibodies appear in the serum. Provided cholera vaccine has not been given previously, this may allow a retrospective diagnosis to be made by serological tests. This has no clinical importance, and is only of use in special epidemiological circumstances.

TREATMENT

Initial rehydration

Rehydration is the mainstay of cholera treatment. In severe cases with hypovolaemic shock, the restoration of blood volume is urgently needed, and this can only be achieved rapidly by intravenous infusion. Because peripheral veins are collapsed in such patients, the initial resuscitative infusion may have to be given via the femoral or subclavian vein in adults or the internal jugular vein in children.

Fluid in the initial stages is run in as quickly as possible: an initial rate of 4 l/h for the first few litres is the norm in adults. The best guide to success is the return of a palpable arterial pulse. As soon as the systolic blood pressure reaches 90 mmHg, renal function usually returns. Tubular necrosis usually develops only if resuscitation is delayed.

In all patients with hypovolaemic shock, the initial fluid deficit will be at least 10% of the body weight. It is a safe rule of thumb to give one-third of the total estimated deficit in the first 20–30 min.

Choice of rehydration fluid

The type of fluid is less important than an adequate quantity. But because patients usually have a metabolic acidosis due to bicarbonate loss, a deficiency of potassium and a loss of water greater than of salts, a slightly hypotonic alkaline fluid enriched with potassium is the most physiological choice.

The single fluid that meets all these needs, and is suitable both for adults and children, is Ringer lactate solution (BP). This contains calcium 2 mmol, chloride 111 mmol, lactate 27 mmol, potassium 5 mmol and sodium 131 mmol/l. It is suitable for both initial rehydration and maintenance therapy. The WHO intravenous diarrhoea treatment solution (glucose 10 g, potassium chloride 1 g, sodium acetate 6.5 g and sodium chloride 4 g in 1 l) contains acetate 50 mmol, chloride 80 mmol, potassium 10 mmol and sodium 120 mmol/l approximately. Simpler solutions can be used with almost as good results, certainly in adults.

Isotonic sodium chloride solution and isotonic sodium bicarbonate, lactate or acetate can be used in the ratio of 2 vol sodium chloride to 1 vol of any of the alkaline solutions. This is done not by mixing the solutions but by giving two bottles of one and one of the other in rotation. As soon as the blood volume has been restored and the pulse has returned, the drip can be moved to a more convenient site, because the peripheral veins will reappear. Vomiting will cease as soon as the acidosis has been corrected, and fluids can now be taken by mouth. If potassium was not replaced by infusion, it can now safely be given orally.

Maintenance hydration

When the patient has been resuscitated, careful charting of fluid intake and output must be started. The uncontrollable watery diarrhoea will often continue for several days and, once adequately rehydrated, daily losses may exceed 20 l. To measure this accurately it is best for the patient to be nursed on a special 'cholera cot' – a frame bed covered in rubber sheeting with a hole in the middle – to allow the fluid escaping from the anus to be funnelled into a calibrated collecting bucket below.

The urine output must also be charted accurately, and intravenous fluid input should equal the combined volume of stool and urine, plus 500 ml added for insensible losses. Any inaccuracies in the calculations are usually adequately compensated for by the patient's new-found ability to drink. This period of maintenance parenteral fluid therapy can be greatly shortened by the early resumption of oral rehydration with a sugar–electrolyte solution and tetracycline by mouth.

Oral rehydration with glucose–electrolyte solution

This is used for maintenance hydration in severe cases requiring intravenous therapy for resuscitation, and for all milder cases from the beginning. It is much cheaper than intravenous therapy, requires no special apparatus or skills, and is free from the dangers of fluid overload. Its success depends on the fact that glucose crossing into the blood from the gut lumen takes electrolytes with it. If glucose is not available, sucrose can be used with almost as good results, as it is rapidly split into glucose and fructose by intestinal enzymes. The secret of successful oral rehydration is to give the fluid frequently, but in small amounts. In developing countries where nursing resources are strained, the task of oral rehydration is often delegated to relatives, especially the mothers of small children. The following solution is suitable for oral rehydration of adults and children, and special spoons for measuring out the various ingredients are available.*

In 1 l of sterile water: dextrose (glucose) 20 g, potassium chloride 1.5 g, sodium bicarbonate 2.5 g and sodium chloride 3.5 g.

Composition:
Bicarbonate 30 mmol, chloride 80 mmol, potassium 20 mmol and sodium 90 mmol/l.

If glucose is not available, sucrose can be used instead, but should be increased to 40 g/l, as it generates only half its weight of glucose on hydrolysis. Formulas using locally available carbohydrates such as rice powder instead of sugar have been used successfully in many areas.

Various prepackaged commercial preparations are available in sachets and have obvious advantages.

ADMINISTRATION OF ORAL REHYDRATION SOLUTION

This is either by drinking two to four times an hour or by nasogastric tube. The dosage varies with the calculated deficit and rate of fluid loss. In severe continuing diarrhoea it is 15 ml/kg per h, in frequent divided

* Teaching Aids at Low Cost (TALC), Institute of Child Health, 30 Guildford Street, London WC1N 1EH.

doses. If this rate of oral administration cannot keep pace with fluid loss (equivalent to 20 l/day for an adult), parenteral therapy is needed. For mild to moderate diarrhoea a dose of 5–10 ml/kg per h will be adequate.

Tetracycline in cholera

Tetracycline, or any of its relatively non-absorbed derivatives, such as oxytetracycline and chlortetracycline, is given to cholera patients as soon as vomiting has stopped. The normal adult dose is 500 mg 6-hourly for 3 days. This shortens the duration of the diarrhoea by several days, and the stools usually become free of vibrios within 24–48 h, in contrast to 7 days in untreated cases. Furazolidone 400 mg daily for 3 days is equally effective, as is a single dose of 300 mg doxycycline. There are early reports that chlorpromazine by its effect on cyclic adenosine monophosphate may usefully reduce fluid loss.

EPIDEMIOLOGY

The most recent (seventh) pandemic started in Sulawesi in 1961 and spread relentlessly through the western Pacific, South-east Asia, the Middle East, Africa, eastern Europe and the southern Americas. The organism is the El Tor biotype, and it has effectively displaced classical cholera even from its heartland in the great river basins of India. It differs from classical cholera in two important ways: it more often gives rise to the chronic carrier state; and relatively fewer of those infected develop the classical disease. The chronic carrier state has doubtless facilitated its spread. But there is recent evidence that classical and El Tor cholera harmoniously coexist in parts of the Indian subcontinent. Both classical and El Tor cholera agglutinate 01 antisera. Recent outbreaks of cholera in Bangladesh have been caused by a new strain of a non-01-agglutinating vibrio.

Once an area of low environmental sanitation is struck by El Tor cholera, the disease tends to become endemic due to the presence of chronic carriers. Classical cholera tended to 'move on' and not to recur unless reintroduced. There is more infection by person-to-person contact with the El Tor biotype than with classical cholera, and massive water-borne outbreaks seem less common.

CONTROL OF CHOLERA

In the emergency control of cholera during outbreaks, facilities are required for effective case treatment with adequate supplies of intravenous replacement fluids for severe cases and supplies of oral rehydration solution. Longer-term control depends mainly on improving standards of environmental sanitation. The treatment of carriers and known contacts is relatively unimportant, for despite the repeated intro-duction of cases into western Europe, the disease has never established itself here. Cholera vaccine has no significant part to play in controlling the disease. Vaccination gives perhaps 50% immunity for up to 6 months, but the infection, when it does develop in the vaccinated, is not of reduced severity.

Diarrhoea in general

The main problem with any clinical condition is, What should I do? The answer will usually be obvious if the correct diagnosis is made. In the case of acute diarrhoea, it is often impossible to make an exact aetiological diagnosis. Fortunately, this does not usually matter, for the correct action that should be taken can usually be deduced from other factors. This may seem rather abstract, but we hope it will become clear as we go on.

It is usually possible to make a diagnosis with reasonable accuracy based on very few observations. Among the most important are those which come out in the history. But the history must always contain the answers to specific questions with a high discriminating value. It is best to make sure at the start that the patient's idea of diarrhoea and yours are the same.

HISTORY

After taking the narrative history, ask:

1 How long have you had diarrhoea?
2 Was there (or is there) fever?
3 What is the stool appearance and, in particular, does it contain blood or mucus? (Inspect the stool if possible.)
4 How frequent are the motions?
5 Is there any abdominal pain?
6 Is there any tenesmus? (Cramps in the rectum, often felt immediately after defecation.)
7 Is there any vomiting?
8 Does anyone else have a similar illness that you know of?

Treatment must be based on the clinical diagnosis, for even if available, stool culture results usually come too late to influence management.

Interpretation of the answers

1 *Duration.* We define diarrhoea of more than 2 weeks' duration as chronic, because after this length of time many of the causes of acute diarrhoea can be discarded and a largely different list must be considered.

2 *Fever.* Subjective fever implies an infection, although the temperature may rise with dehydration, whatever its cause. The fever may be due to an infection outside the gut such as malaria and does not necessarily imply an infective enteritis.

3 *Blood.* This usually signifies ulceration of the large bowel. It has a high discriminating value in tropical countries where haemorrhoids are uncommon.

4 *Frequency.* This is a useful guide to the severity of the diarrhoea. Very frequent stools associated with vomiting warn of the likely development of dehydration.

5 *Abdominal pain.* This is most severe in conditions causing inflammation of the gut, such as *Campylobacter* and *Shigella* infections, and also when a necrotizing toxin, such as that of *Clostridium perfringens*, is at work. A proper history should be able to distinguish between these visceral pains and the cramps of the abdominal muscles that occur in severe cholera.

6 *Tenesmus.* This is an indication of inflammation in the rectum and is most often present in inflammatory bowel disease such as shigellosis.

7 *Vomiting.* This indicates systemic intoxication, although it also occurs with the severe metabolic acidosis of cholera. It occurs with preformed toxin food-poisoning and acute gut infections. It may be a prominent symptom of malaria (especially *Plasmodium falciparum*) in non-immunes.

8 *Others infected.* The object is to identify a common source outbreak. This is a much more useful question than 'What have you been eating recently?' which is usually answered by a time-wasting catalogue of scanty relevance.

Simple classification based on three crucial questions

The three crucial questions are: duration? blood? fever? With their help, eight categories of diarrhoea can be rapidly defined — four for acute diarrhoea and four for chronic diarrhoea (Fig. 16.1).

So the categories are:

1 Diarrhoea with fever and blood.
2 Diarrhoea with fever, without blood.
3 Diarrhoea without fever, with blood.
4 Diarrhoea without fever or blood.

This is helpful when considering acute diarrhoea, but less so for chronic diarrhoea.

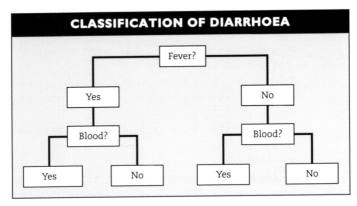

Fig. 16.1 Simple classification of diarrhoea.

ACUTE DIARRHOEA

Acute diarrhoea with fever and blood

The main causes are:
1 Bacillary dysentery.
2 *Campylobacter* enterocolitis.
3 *Salmonella* enterocolitis.
4 Verotoxin-producing strains of *Escherichia coli*.

In the first three cases, the stool will contain an inflammatory exudate, and macrophages are especially common in bacillary dysentery. The inexperienced microscopist often mistakes them for amoebae. Sheets of polymorphonuclear leucocytes in stained smears made from mucus are very characteristic of shigellosis. In all four conditions vomiting is often present.

The distinction between the four conditions cannot be made clinically, although *Campylobacter* infection tends to cause more abdominal pain and runs a longer course than the other three. *Campylobacter* infection is mainly a disease of childhood in developing countries. If dark-field microscopy is available, *Campylobacter* can be identified by an experienced microscopist in at least 80% of wet specimens taken from acute cases. The spirillum-like organisms are very small, and move in a characteristic way. So far there is no specific antiserum available to help with their identification as with cholera.

The likelihood of a *Salmonella* aetiology is greatly increased if the patient is one of many who attended a meal where meat was served. Only a minority of patients with *Salmonella* food poisoning pass blood in the stool, as the main pathology is usually in the small bowel. Verotoxin-producing *E. coli* infection is sometimes complicated by the haemolytic–uraemic syndrome.

A PROBLEM OF MANAGEMENT

There has been an outbreak of bacillary dysentery in a refugee camp in Thailand. As the medical officer in charge, which of the following treatment policies would you adopt?

1 Treat all patients and close contacts with a 5-day course of tetracycline.

2 Treat only the patients with tetracycline.

3 Organize an emergency treatment centre where all dehydrated patients receive intravenous fluids.

4 Treat all patients with ciprofloxacin.

5 Organize an emergency treatment centre to provide intravenous rehydration for severe cases, oral rehydration for less severe cases, and selective chemotherapy.

The only sensible answer is (5).

Resistance of *Shigella* to antibiotics is now worldwide, and in few places more pervasive than in South-east Asia. Much of this is due to plasmid-mediated transferable resistance, acquired from the normal coliform inhabitants of the bowel. It has been encouraged by the widespread use of antibiotics for trivial conditions, the unethical activities of some drug companies, and the free availability of antibiotics 'over the counter' in many parts of the world.

Shigella dysentery is usually self-limiting, and no matter which organism is responsible, supportive therapy with appropriate fluid and electrolyte replacement (see Chapter 15) is usually all that is needed. In very toxic cases, an antibiotic may be used as well, in the hope that the organism is sensitive.

Among the common ones most likely to work are co-trimoxazole, ampicillin or amoxycillin, and nalidixic acid. One of the newer quinolone antibiotics such as ciprofloxacin would be a good choice if it can be afforded. Since antibiotic resistance patterns are constantly changing it is wise to find out the current resistance pattern from your nearest microbiological laboratory.

As *Salmonella* enteritis usually presents as diarrhoea without blood, it appears in the next section.

CAMPYLOBACTER ENTERITIS

This is now recognized as the commonest infective cause of bloody diarrhoea in the UK. Severe upper abdominal pain may lead to the patient's admission as a surgical emergency. Many patients need parenteral rehydration, because of the persistent vomiting. The illness is shortened by a suitable antibiotic. Although ciprofloxacin or erythromycin are the drugs of choice, there is usually a satisfactory clinical response to co-trimoxazole. This simplifies the management of infective febrile diarrhoea: if an antibiotic is indicated (and it usually is not) then co-trimoxazole or ciprofloxacin can be used in all cases.

Organisms which cannot be recognized on dark-field microscopy, can be grown under microaerophilic conditions on a special culture medium. In developing countries the infection is usually not identified.

Some textbooks say that falciparum malaria can cause bloody diarrhoea. If so, it must be very rare. The frequent occurrence of bacillary dysentery in visceral leishmaniasis is worth remembering.

Acute diarrhoea with fever but no blood

The main causes are:
1 *Salmonella* enteritis.
2 Malaria, especially *P. falciparum*.
3 Almost any infection in a child.
4 Mild (non-ulcerating) shigellosis.
5 *Campylobacter* infections.

In a busy clinic in a developing country, one usually has little time to examine patients with diarrhoea, apart from making a rapid assessment of hydration and toxaemia. But one group should be carefully examined: children with diarrhoea and fever, because the diarrhoea might be a manifestation of infection elsewhere, detectable only on careful examination. These infections include otitis media, tonsillitis, pneumonia and urinary tract infection, and a blood film for malaria should always be taken.

SALMONELLA ENTERITIS

This tends to occur in food-borne outbreaks. It is not usually spread by infected water supplies because the dose needed to establish infection is so large. For the same reason, person-to-person spread is rare (contrast with *Shigella*).

Normally, the organisms need to multiply in infected food before achieving a concentration sufficient to ensure human infection. Infected food looks and tastes wholesome.

The onset is typically with fever and chills, followed by vomiting and diarrhoea. The incubation period, usually 12–48 h, reflects the phase of multiplication in the gut. More than 100 named *Salmonella* species can cause the syndrome. The infection is a zoonosis, the natural reservoirs being in animals, including mammals, reptiles and birds.

Invasive salmonellosis giving a typhoid-like picture is especially common in HIV-infected patients.

ANOTHER PROBLEM WITH DIARRHOEA

The first patient to attend your morning clinic is a bride, whose wedding took place 2 days ago. For 12 h she has had fever and vomiting, followed later by diarrhoea. There are more than 20 of the wedding guests outside your clinic. It is obvious that many of them are suffering from similar symptoms. Cold chicken was served at the wedding. The predominant syndrome is: diarrhoea, abdominal cramps, vomiting and fever. You make a clinical diagnosis of *Salmonella* gastroenteritis. How should you treat the affected patients?

Treatment of *Salmonella* gastroenteritis

The mainstay of treatment is supportive and directed towards re-hydration (details in Chapter 15, under Treatment). The condition is usually self-limiting, and spontaneous recovery normally occurs in a few days. Therefore antibiotic treatment should not usually be given. However, some patients do need an antibiotic and must be identified.

Indications for antibiotic treatment of salmonellosis

The main indication is when the disease becomes invasive: the bacteria have then spread to the blood stream and the illness is now as serious as typhoid. The main indications of invasive disease are:

1 Fever of more than 48 h duration (in an adult).
2 The appearance of rose spots (reddish macules scattered over the trunk).
3 Splenomegaly.

In very sick, small children who look toxic, antibiotic therapy will often be started when the child is first seen, on the grounds of informed clinical guess. If one waited for 48 h in these cases, some children would die. Chemotherapy should be given just as for typhoid and continued for 14 days to prevent relapse. Relapsing *Salmonella* septicaemia is sometimes associated with *Schistosoma mansoni* infection.

Antibiotic treatment should be given to children with sickle-cell disease who develop *Salmonella* enteritis, to protect them from *Salmonella* osteomyelitis. It should also be given to those known to be HIV-infected since the risk of invasive disease is greatly increased.

Diarrhoea without fever but with blood

Much the most important cause of this is amoebiasis but, less commonly, it is due to *Balantidium coli* infection. Acute *Schistosoma mansoni* or *S. japonicum* infections may cause this (eosinophilia is a suggestive feature) and in the early stages there may be some fever. Very severe bloody diarrhoea may develop when colonic schistosomal pseudopolypi ulcerate.

In small children who eat dirt, or in mental defectives in institutions, massive *Trichuris* infections can cause bloody diarrhoea.

Ulcerative colitis is a common cause of this clinical picture in Europe and North America. It is rare in most developing countries, where the diagnosis can only be made by exclusion. This is very important, for corticosteroid treatment of unrecognized amoebiasis would be disastrous.

Pseudomembranous colitis is caused by certain strains of *Clostridium difficile*, almost always following antibiotic treatment.

Acute diarrhoea without blood or fever

The main causes are preformed toxins from food poisoning, or colonization of the gut by organisms which make toxins *in situ*. Occasionally, diarrhoea is due to simple chemical components of foods.

A careful history usually gives vital clues to the likely aetiology. The following causes are the commonest:

1 Enterotoxin-producing strains of *Staphylococcus aureus* (staphylococcal food poisoning).
2 Enterotoxigenic strains of *E. coli* (travellers' diarrhoea).
3 *Clostridium perfringens* (clostridial food poisoning).
4 Viral infections of the gut (most important are the rotavirus infections of early childhood).
5 Food toxicants.

ENTEROTOXIN-PRODUCING STRAINS OF STAPHYLOCOCCUS AUREUS

These can grow in a variety of prepared foods, following inoculation usually from the hands of an infected food-handler, but sometimes from the milk of cows with staphylococcal mastitis. Because there is a preformed toxin, the bacteria do not need to multiply in the host, so the incubation period is very short, usually 2–6 h. Vomiting is soon followed by diarrhoea, which may be almost as severe as that of cholera. There is never any colonic ulceration, so blood is not seen in the stool.

The illness is very brief, terminating as soon as the toxin is eliminated. But it may be very severe, and DRB observed 2 deaths in a group of 40 elderly north American matrons infected on an aeroplane *en route* for Bangkok. Treatment, although purely supportive, may need to be vigorous. Obviously, antibiotics are useless.

ENTEROTOXIGENIC ESCHERICHIA COLI

This is probably the most important single cause of travellers' diarrhoea, but toxin production is not the only factor. Other mechanisms include the ability to adhere to and invade the small intestinal mucosa. Clinical recovery is followed by prolonged immunity, for which reason this is mainly a disease of visitors rather than residents. It may be a fairly important cause of childhood gastroenteritis in developing countries, and there are many recorded outbreaks among children in institutions in Europe and north America. Specific diagnosis requires help from a technically advanced laboratory. Fortunately, as treatment is supportive only, failure to recognize the exact aetiology is not very important. Once tolerance to the pathogenic effects of infection is acquired, the organisms become part of the normal colonic flora. The fact that all long-stay residents of endemic areas harbour the organisms must largely account for the high attack rate in visitors. The local name for the condition often reflects the rueful humour of its victims—Hong Kong dog; Delhi belly; gippy tummy; Montezuma's revenge; the Aztec two-step.

The diagnosis is usually clinical. Treatment is supportive. Antibiotics are often ineffective, unless one of the new quinolones is used, because of the widespread nature of plasmid-mediated transferable resistance. Drugs which offer symptomatic relief by reducing intestinal mobility such as loperamide (Imodium) are useful in adults for social convenience but they do not treat the infection and should be avoided in children and when diarrhoea is bloody. There is good evidence that such drugs delay the final elimination of the offending organism.

There is some evidence that attacks of travellers' diarrhoea can be reduced in frequency by prophylactic treatment with bismuth sub-salicylate 2 × 262 mg tablets chewed well four times a day—an inconvenient drug which is not available in the UK. Antibiotics should rarely be used for prophylaxis, both because of the risks of side-effects and the encouragement of drug resistance. Travellers who are regarded as at particularly high risk because of other medical conditions can be protected by one of the quinolone antibiotics such as ciprofloxacin 500 mg daily.

Travellers' diarrhoea can often be prevented by strict adherence to the adage: 'Cook it, peel it or leave it'.

CLOSTRIDIAL FOOD POISONING

This is due to the ingestion of food massively contaminated with *Clostridium perfringens*, which multiplies in the food, especially meat, at ambient temperatures. Contamination with spores which are relatively resistant to heat accounts for infection in meat which appears fairly well cooked. The type A strains cause diarrhoea with abdominal cramps due to enterotoxin. Type C strains may invade the gut mucosa and lead to necrosis of the full thickness of the intestinal wall. This most commonly follows communal feasting on an infected carcass. In Papua New Guinea, the association with pig feasts is so well known that the condition is called pig bel. Severe cases present with abdominal pain and ileus, rather than with diarrhoea. When peritonitis and signs of intestinal obstruction are present, surgery offers the only (albeit slim) hope. A vaccine is available against type C toxin.

VIRAL CAUSES OF DIARRHOEA

The viruses with a long-established affinity for the gut mucosa, the enteroviruses, do not seem to be a very important cause of diarrhoea. For example, the virus of poliomyelitis does not usually cause diarrhoea at all.

Rotavirus infection is an important cause of gastroenteritis in children throughout the world. The failure to make the specific diagnosis is of no account, because, as in most other acute diarrhoeal conditions, the treatment is supportive only.

FOOD TOXICANTS AND DIARRHOEA

Although there are some foods which commonly cause diarrhoea in those who are not used to them, they are not a common cause of diarrhoea in expatriates taking up residence in the tropics. Much more common is enterotoxigenic *E. coli* infection.

Food toxicant diarrhoea has a very short incubation period, and is short-lived. The best-known toxin is capsaicin, the substance that makes chillies hot. Taken in excess, it can create a biochemical lesion in the small bowel like that of cholera. The distinctive symptom is the perianal burning that follows defecation. The Indians say the 'chilli is tasted twice'. No treatment is needed. Prevention is a matter of common sense.

PERSISTENT DIARRHOEA

Following an attack of what sounds from the history to have been acute infective diarrhoea, many patients continue to have diarrhoea of lesser severity for weeks or months. There are five main causes:
1 Secondary disaccharidase deficiency.
2 Concomitant infection with another organism such as *Giardia lamblia* or *Entamoeba histolytica*.
3 Tropical sprue.

4 HIV infection, especially if there is a failure to clear pathogens such as *Cryptosporidium* or *Isospora* species.

5 Infection with an enteropathogenic *Escherichia coli*, causing long-lasting damage to the brush border of the enterocytes, particularly in children.

Lactose intolerance: a typical history

A chemistry graduate and his girlfriend had recently returned to the UK from Nepal. Both had developed diarrhoea (without fever or blood) soon after reaching Kabul. The diarrhoea had been pale, frothy and foul-smelling. Both had lost about 7 kg in weight.

Stool examination in our laboratory revealed cysts of *G. lamblia*. Both were treated with metronidazole. The girl recovered completely. The young man improved, but continued to pass several loose motions every day. I told him I thought it most likely that he had secondary lactase deficiency causing lactose intolerance, especially as he was a lactovegetarian. He rejected my advice to stop drinking milk as unscientific. He was given 50 g of lactose by mouth the next morning, and had his blood glucose level estimated every 30 min thereafter. Half an hour after the lactose he developed severe abdominal pain, and profuse watery diarrhoea followed soon after. The blood glucose levels showed no significant rise. After withdrawing milk and milk products (even yoghurt contains residual lactose after fermentation), his diarrhoea cleared up completely. A month later milk was gradually reintroduced to the diet without further trouble.

Reducing substances, commonly present in the stools of children with lactose deficiency, are not usually found in the stools of adult patients.

When advising the patient to avoid lactose, remember that lactose is a commonly used excipient in various tablets.

Failure of diarrhoea to resolve after a bowel infection may be the way chronic diarrhoeal conditions such as tropical sprue or HIV infection present. An episode of acute infective diarrhoea often precedes the onset of ulcerative colitis.

FURTHER READING

DuPont, H.L., Ericsson, C.D. (1993) Prevention and treatment of traveler's diarrhea. *N Engl J Med* **328**, 1821–7.

Diarrhoeal disease: curent concepts and future challenges. (1993) *Trans R Soc Trop Med Hyg* **87** (suppl 3), S1–53.

SECTION 3

Soil-transmitted helminths

These are all nematode infections of the bowel, which are spread from human to human. The one exception is the zoonosis–toxocariasis. The egg or larva of the worm is not infective when first passed in the stool, but has to undergo further development in the soil, so transmission can only occur when the ground is contaminated with faeces.

Ascaris lumbricoides: the large round worm

Adults are large, cream-coloured worms; the males are 15–30 cm long, females 20–40 cm. They live in the small intestine and obtain nourishment from the intestinal contents. They do not suck blood or damage the mucosa significantly. About 1 in 4 of the world's population is infected, and the infection is found worldwide where conditions of environmental hygiene are low. Because of their dirtier habits, children are more often, and more heavily, infected than adults.

LIFE CYCLE

The fertilized female lays about 200 000 eggs a day. These pass out in the stool and in favourable conditions develop into the infective (embryonated) stage in 10 days. Eggs remain infective for weeks or months or even years, but are destroyed by direct sunlight. When the embryonated egg is swallowed, it hatches in the intestine and liberates a larva. This migrates through the intestinal wall and via the circulation reaches the lungs. Further growth and development occur in the lung and finally the larva penetrates the alveolar wall and is transported, via the ciliary escalator, to the epiglottis. It is then swallowed to reach its final habitat. From the time the egg is swallowed to the time the worms become mature is 60–75 days.

The morphologically similar *Ascaris* of pigs, *A. suum*, also infects humans.

SYMPTOMS DUE TO PULMONARY MIGRATION

Ascaris larvae are large and antigenic and cause damage during their entry into the alveoli. Symptoms may be severe if many larvae are migrating at the same time. The condition is then called verminous pneumonia. The clinical picture is of cough, fever, wheeze, dyspnoea in severe cases, rapidly changing alveolar X-ray shadows and eosinophilia. This is Loeffler's syndrome. There may be eosinophils in the sputum. The condition subsides spontaneously in a few days—longer if successive waves of larvae are passing through.

EFFECTS OF ADULT WORMS

As in all worm infections where the worms do not multiply in the body, the effects depend on the worm load. The more numerous the worms, the greater the chances of severe disease. Most infections are asymptomatic. The main effects are:

1 Mechanical.
2 Toxic.
3 Metabolic.

Mechanical effects

1 In the small bowel: impaction may cause obstruction, volvulus or intussusception.
2 In the appendix: appendicitis.
3 In the biliary tree: obstructive jaundice and intrahepatic abscesses.
4 In the pancreatic duct: pancreatitis.
5 In the larynx: worms vomited up may occasionally cause asphyxia.

Toxic and metabolic effects

These are vague and ill-understood but a heavy worm burden may significantly contribute to malnutrition.

PRESENTATION OF ASCARIASIS

1 Mild bouts of recurrent colic.
2 The mother has seen a worm. It may be passed in the stool or vomited up, or even emerge from the ear via a perforated ear drum.
3 The child may present with complications.
4 The infection may be suspected in a poorly nourished, pot-bellied child with oedema.

Worms tend to be expelled during bouts of diarrhoea of any cause. This does not mean that the worms cause the diarrhoea.

ASCARIS AND THE SURGEON

Intestinal obstruction usually only occurs in children, and intrahepatic worms are also typically a childhood complication.

If the intestine is opened for any reason in a patient with ascariasis, the worms will very likely breach the suture line and enter the peritoneum. When bowel surgery is carried out, local worms should be removed, and piperazine suspension in normal doses injected upwards and downwards to prevent this happening.

Even if escaping worms do not cause an obvious pyogenic peritonitis, they may continue to lay eggs and give rise to a granulomatous peritonitis due to granulomata developing around the eggs.

DIAGNOSIS

Direct diagnosis

This is by finding the characteristic eggs in the faeces. As each worm lays about 200 000 eggs/day, they are easily found on the direct-smear examination. There is no need for a concentration technique. In most symptomatic cases there are enormous numbers of eggs, which can be found within a few seconds of starting to scan the slide.

Indirect diagnosis

Immunodiagnostic methods do exist. They are of no clinical importance.

Circumstantial diagnosis

Apart from mild eosinophilia, there are no distinctive findings.

TREATMENT

Piperazine salts

This treatment is the most widely used. It is available as tablets and suspensions of various strengths. The dose is 75 mg/kg as the hydrate (maximum 4 g) on 2 successive days (100 mg hydrate = 125 mg citrate, 120 mg adipate, 104 mg phosphate). Two doses give a cure rate of over 90%. Toxic effects are usually mild: nausea, vomiting, ataxia ('worm wobble').

Pyrantel embonate

A single dose should be given: 10 mg/kg. It is more effective than a single dose of piperazine and has useful activity against hookworms too. There are mild toxic effects (headache, dizziness, vomiting, abdominal pain, diarrhoea) in 20% of patients.

Mebendazole

The dose is 100 mg twice daily for 3 days for all ages over 2 years. It is contraindicated in pregnancy. It is active against hookworm and *Trichuris* also and is very well tolerated.

Levamisole

A single dose of 2.5 mg/kg should be given. It is very effective and there are toxic effects in fewer than 1% of cases.

Albendazole

This is a broad-spectrum anthelmintic effective in a single dose of 400 mg for adults and children over 2 years old. Below 2 years, give a dose of 200 mg. It is also active against *Trichuris* and hookworms. It is contraindicated in pregnancy.

Which drug is best?

In developing countries the deciding factor is often cost. The cost of an average adult dose of levamisole should not be much more than £0.10.

EPIDEMIOLOGY AND CONTROL

Disease is chiefly in children and they usually contaminate the environment most. Periodic mass treatment of children will eliminate most disease. Transmission will not be interrupted unless chemotherapy is combined with health education and high levels of environmental sanitation.

FURTHER READING

Crompton, D.W.T., Nesheim, M.C., Pawlowski, Z.S. (eds) (1989) *Ascariasis and its Prevention and Control.* Taylor & Francis, London.

CHAPTER 18

Hookworm

The hookworm is widespread in the tropics and subtropics and can even persist in north Europe if a suitable microclimate exists, such as in mines. The worms inhabit the small intestine of humans and suck blood. Adult worms live several years. There are two major hookworms in humans:

1 *Ancylostoma duodenale*, the main hookworm in south Europe, north Africa, north India, north China, Japan and the west coast of south America; it is also, together with *Necator americanus*, found in large areas of South-east Asia, the Pacific and west Africa.

2 *Necator americanus*, the main hookworm of sub-Saharan Africa, south Asia and the Pacific; is also widespread in the Americas.

ANATOMY

Hookworms are slender tubes about 1 cm long. They have a mouth and oesophagus at the front, connected by the gut to the anus at the rear. *Ancylostoma* is larger than *Necator*. The female body is largely occupied by eggs. The mouth and pharynx are used to attach the worms to the mucosa by suction. The teeth in *A. duodenale* and the cutting plates in *N. americanus* are used to pierce the mucosa. Their jaws are rigidly fixed in a gape.

LIFE CYCLE

Fertilized females lay eggs which pass into the bowel lumen and leave the body in the faeces. Freshly passed eggs contain a few large cells. In warm, moist surroundings, the first-stage larva emerges in less than 24 h, and infective third-stage (filariform) larvae develop in a week. They remain infective in the soil for many weeks or months. These filariform larvae normally infect by penetrating the intact skin, but *A. duodenale* larvae, when swallowed, can pass directly to the gut mucosa.

Fate of the larva

1 The larva migrates from the skin to the lung on the third day (via the lymphatic system and blood stream).
2 Then follows penetration of the capillary wall to enter the alveolus.
3 Then the larva migrates via ciliary propulsion up the respiratory tree to the epiglottis.
4 It is then swallowed.
5 It reaches the intestine in 1 week, is fully grown in 2–3 weeks, and sexually mature in 3–5 weeks (eggs then begin to appear in the stools).

A. duodenale, after the initial human infection, may enter a dormant phase of arrested development lasting many months, leading to a prolonged prepatent period. This is believed to be due to evolutionary adaptation to unfavourable environmental conditions such as a long dry season which would otherwise be inimical to its chances of survival. The phenomenon is called hypobiosis.

CLINICAL EFFECTS OF HOOKWORM INFECTION

1 At the site of larval penetration: sometimes some itchy papules; this is ground itch.
2 During invasion of the circulation: usually no ill-effects.
3 During migration through the lungs: sometimes cough, wheeze, transient X-ray shadows. The symptoms are not severe (contrast with *Ascaris*).
4 In the gut: the only important clinical effects are due to losses of iron and proteins into the gut, mostly from the anus of the worm.

FACTORS GOVERNING DEVELOPMENT OF ANAEMIA

1 Number of worms (the greater the worm load, the greater the blood loss). About half the iron expelled by the worms is reabsorbed. Each *N. americanus* causes the loss of about 0.05 ml/day.
2 Species of worm (*A. duodenale* consumes four to five times as much blood a day as does *N. americanus*).
3 Iron intake (this, and the worm load, are the most important factors. Iron intake varies between 4 and 40 mg/day in different parts of the world).
4 Availability of iron (in some tropical diets, a high phytic acid intake reduces the availability of iron).
5 Iron absorption (the worms themselves do not seem to cause malabsorption, but something else might).

6 The presence of other conditions increasing iron demand, such as pregnancy and menstruation.

HOOKWORM ANAEMIA (HWA)

This is usually pure iron deficiency in type. Occasionally associated folate deficiency is revealed when iron is replaced. Anaemia does not develop until iron reserves are exhausted, after a prolonged period of negative iron balance. In areas of high hookworm prevalence, all anaemic patients will have hookworms. However, it does not follow that all these patients have HWA.

Symptoms

1 Tiredness.
2 Aching muscles.
3 Breathlessness.
4 Oedema.
5 Pallor.
6 Dysphagia—very rare.

Signs

1 Pallor of skin and mucous membranes (in dark-skinned people, some of the pallor is due to melanin depletion).
2 Oedema (may be related to heart failure or low serum albumin levels).
3 Koilonychia. (This is only significant if present in the fingers; people who habitually walk barefoot often have koilonychia of the toes without anaemia. The cause is unknown.)
4 A smooth red tongue and angular stomatitis are very rarely seen.

Natural history of untreated HWA

In some cases the condition stabilizes when the haemoglobin (Hb) content falls to a certain level. This is because the iron loss caused by the loss of blood falls as the Hb content (and so the iron content) of the blood falls. So:

y ml of blood lost per day, Hb 16 g/dl, iron loss from $16 \times y$ ml of blood lost per day; later Hb 8 g/dl (the patient is now anaemic), iron loss from $8 \times y$ ml per day.

At the lower level of loss, the iron content of the diet may be adequate to maintain the lower Hb level. In most hookworm infections, the increased iron loss caused by the worms can be fully compensated for by increased iron absorption from the diet, which in most parts of the world contains more iron than is needed to meet normal demands. So most hookworm infections merely cause an increase in absorption of dietary iron and no anaemia.

The worm load tolerated without developing anaemia varies greatly in different parts of the world. Albumin loss is only significant in very heavy infections.

DIAGNOSIS

This is by finding the characteristic eggs in the stool. If the specimen is kept at ambient tropical temperatures for many hours, the eggs may hatch and liberate the rhabditiform larvae. These may easily be confused with larvae of *Strongyloides stercoralis*, for which reason stool specimens should be examined promptly. A direct-smear examination is adequate because, if there are so few eggs that a concentration technique is needed to find them, the infection is of no clinical significance. Apart from the exceptions listed under Paradoxes, see p. 190, if no hookworm eggs can be found, one can forget about hookworm as a cause of the patient's symptoms. There is no convincing evidence that hookworm infection causes significant intestinal pathology (apart from blood loss) or symptoms unrelated to anaemia.

If hookworm infection is light, and not causing anaemia, there is no need to treat it. To do so in an endemic area would be to waste scarce resources to no effect, unless mass treatment were to form part of a control programme.

FALSE HOOKWORM

Humans may be infected with largely non-pathogenic worms whose eggs resemble those of hookworm. The most important is *Ternidens deminutus*, a common parasite in monkeys, baboons and humans in southern Africa. The worms inhabit the large bowel, where they may cause cystic nodules. Because they suck blood, they can cause anaemia in heavy infections. Their eggs closely resemble those of the hookworm, but are larger. Failure to distinguish these worms from true hookworm has led to great confusion in drug trials carried out in southern Africa, for *Ternidens* is much more susceptible to certain drugs than are the true hookworms.

Trichostrongylus worms of many species are natural parasites of herbivores in many parts of the world, and humans can become infected by ingesting the infective larvae with vegetables or salads. The adult worms are attached to the small intestine, but cause only mild damage and insignificant blood loss. The eggs of *Trichostrongylus* spp. have more pointed ends than the eggs of true hookworms.

CULTURING NEMATODE EGGS IN FAECES

If eggs passed in the faeces are allowed to develop in favourable conditions, they will eventually develop into infective larvae which show clearly discernible differences of morphology. The simplest method is the Harada–Mori technique, in which faecal material is smeared on to a slip of blotting paper, one end of which is allowed to rest in a small puddle of water in a tubular plastic bag. After 7–14 days, infective (identifiable) larvae will have developed and fallen into the water at the bottom of the bag. The larvae can be aspirated, killed and stained with iodine, and examined under the microscope. This method allows false hookworms to be distinguished from true hookworms and also enables one to distinguish between *A. duodenale* and *N. americanus* infections. The alternative method is to 'worm' the patient (actually deworm), administer a purge, and pass the resulting faeces through a sieve. The adult worms possess distinctive features which are described in parasitology textbooks. The culture method is less unpleasant, both for the patient and for the investigator.

DIAGNOSING HWA

There are three stages:
1 Demonstrate iron-deficiency anaemia.
2 Show the presence of significant numbers of hookworm eggs in the stools.
3 Exclude other causes of iron loss.

For diagnosis of iron deficiency, see Section 7. To find significant numbers of hookworm eggs in the stool depends on local knowledge of the relationship between hookworm load and iron intake. In west Africa, hookworm egg outputs of less than 10 000/g are unlikely to be important because of the high dietary iron intake. This corresponds to 10 hookworm eggs/mg of stool, and as the normal direct smear contains 2–4 mg, the hookworm egg count per cover-slip preparation would be at least 20–40. But in areas of low iron intake, a tenth of this count may be highly significant.

Whatever method is used, some estimate of egg count (and hookworm load) must be made. The most accurate method is by dilution of a known amount of stool, as in the Stoll count (consult a parasitology textbook). But the method is time-consuming and, for ordinary clinical purposes, it is enough to ask the laboratory to report the number of hookworm eggs per slide. Experience will soon show what counts are likely to be associated with anaemia.

Excluding other causes of iron loss is largely a matter of taking a good history with special attention to menstrual losses and pregnancies in women, and gastrointestinal bleeding in both sexes. Do not forget that thalassaemia and prolonged severe infections can both mimic iron deficiency.

Indirect and circumstantial diagnosis of HWA

Immunodiagnosis is clinically useless in diagnosing HWA. Circumstantial diagnosis (where the methods are available) would include faecal occult blood estimation, serum iron and ferritin levels. Eosinophilia, such a common accompaniment of helminth infections, tends to disappear when HWA develops. In heavy infections (mainly in areas of high iron intake) albumin loss may lead to hypoalbuminaemia.

Paradoxes in HWA

1 The patient has iron-deficiency anaemia but no hookworm eggs in the stool. (This situation occurs when a patient with HWA is treated with an anthelmintic alone, but is not given iron.)
2 The patient has a very heavy hookworm infection but is not anaemic. (The infection is recent, and has not been present for long enough to exhaust the iron stores.)
3 There is no correlation in the community between anaemia and hookworm infection. (Infection is not the cause of anaemia; *heavy* infection is. The investigator has not bothered to count the hookworm eggs. This explains much of the early confusion about the cause of HWA.)

TREATMENT OF HWA

Replacing iron

This is the most urgent measure. Replacement should correct not only the iron deficiency in the blood, but also replenish the depleted iron stores.

ORAL IRON

Ferrous sulphate is the drug of choice. The adult dose is 200–400 mg three times daily.

Ascorbic acid, 50–100 mg per dose, increases absorption but is not economically worthwhile in poor countries. Oral iron should be continued for about 3 months after the Hb is normal, to replenish stores.

None of the more expensive iron preparations (ferrous gluconate, ferrous fumarate, ferrous succinate) is more effective than ferrous sulphate in doses containing equivalent amounts of elemental iron. All iron salts seem to cause gastrointestinal side-effects, mainly constipation, in large doses. All are intensely toxic to children when given in overdose. Patients should be warned of this.

PARENTERAL IRON

This is always more expensive than oral iron. Because the Hb level rises no faster with parenteral iron than with properly supervised oral iron (about 0.1–0.15 g/day), parenteral iron should only be used under two circumstances:

1 If the patient cannot be trusted to take oral iron, judged usually from past experience.

2 If the patient is iron-deficient late in pregnancy, and one cannot take the risk that she may not take the tablets. Do not give parenteral iron in the first trimester.

Because intramuscular iron requires repeated injections, the patient must reattend on several occasions. If patients cannot be trusted to take the iron by mouth, in our experience they cannot be trusted to reattend for repeated injections either. For this reason, unless parenteral iron has to be given because of malabsorption (such as in coeliac disease), parenteral iron is best given intravenously by total-dose infusion. The only preparation suitable for this is iron dextran (Imferon), and the precautions described in the package must be strictly obeyed if alarming anaphylactoid reactions are to be avoided.

Getting rid of the worms

In heavy infections, repeated courses of an anthelmintic drug may be needed to achieve parasitological cure. But this is not usually necessary, provided the worm burden is reduced below the level which causes negative iron balance. The choice of anthelmintic depends not only on effectiveness but on tolerability, activity against other worms, relative effectiveness against the hookworm species involved, availability and cost. The cost factor is often as important as any other in developing countries.

TETRACHLOROETHYLENE (TCE)

0.1 mg/kg (maximum 5 ml). Take on an empty stomach with water only for 3 h afterwards. No purge is needed. Several courses may be needed in heavy infections. It is perhaps more effective against *N. americanus* than *A. duodenale*.

Toxic effects

Nausea, vomiting, abdominal pain and intoxication resembling drunkenness.

Precautions

Treat *Ascaris* first, or mass migration may cause intestinal obstruction (this is only important in heavy *Ascaris* infections).

BEPHENIUM HYDROXYNAPHTHOATE (ALCOPAR)

5 g of salt, taken as for TCE. It is relatively ineffective against *N. americanus*, for which many courses may be needed to reduce worm load to acceptable levels. It possesses a useful activity against *Ascaris*. It is safe in pregnancy. It is 20 times as expensive as TCE.

Toxic effects

Gastrointestinal only. It tastes bad.

PYRANTEL EMBONATE (COMBANTRIN)

10–20 mg/kg as a single dose. This is effective against *Ascaris* also.

Toxic effects

See Chapter 17, p. 183.

MEBENDAZOLE (VERMOX)

See Chapter 17, p. 184. This has a useful broad spectrum (hookworm, *Ascaris*, *Trichuris*) and can be obtained cheaply. It is very well tolerated. It is contraindicated in pregnancy.

LEVAMISOLE

2.5 mg/kg as a single dose. This is very effective against *Ancylostoma duodenale* and *Ascaris*. Its effectiveness against *N. americanus* is less well documented. It has a very low toxicity and low cost.

ALBENDAZOLE

400 mg as a single dose (for adults). This is probably the most effective single-dose drug, and its cost is moderate. Available as 200 mg tablets under the name Zentel. Same broad spectrum as mebendazole. Avoid use in pregnancy.

Failure of anaemia to respond to iron

The main causes are:

1 Non-compliance—the patient will not take the medicine. It may be given to relatives or friends, or offered for sale in the market.

2 Concomitant folic acid deficiency—this may be unmasked by replenishing the iron stores. A dimorphic blood picture may be present initially. A megaloblastic picture develops later.

3 Failure to absorb iron—this is rare. The main cause is a small bowel defect, as in coeliac disease.

4 Another unrelated cause of anaemia is present at the same time, such as a bleeding duodenal ulcer or TB.

5 The initial diagnosis is wrong.

EPIDEMIOLOGY AND CONTROL

The epidemiology varies greatly in different parts of the world. Where the family dwelling has no latrine, infection may be most severe in the women and children, who acquire infection near the house. Where there is a household latrine, those most at risk may be the farmers who defecate on their farm and build up a large population of infective larvae in the soil, so creating a positive-feedback loop. Rapid relapse after cure is sometimes due to this. Control cannot be achieved by mass treatment alone, although a great initial effect on worm burden is produced.

A lasting result must depend on health education and an improvement in environmental sanitation. The wider use of protective footwear should reduce transmission, but sandal-type 'flip-flops' suitable from the standpoint of comfort have not proved to be effective.

Where the dietary iron intake and worm burden are both low, iron supplementation of the diet such as by the enrichment of flour may be the most cost-effective measure.

FURTHER READING

WHO (1987) *Prevention and Control of Intestinal Parasitic Infections.* Technical Report Series no 749 WHO, Geneva.

Trichuris (Trichocephalus) trichiura: *the whip worm*

Trichuriasis is a bowel infection of cosmopolitan distribution, most prevalent in warm, humid climates. Symptomatic infection is usually only recognized in small children. Pigs harbour an apparently identical parasite which can infect humans.

LIFE CYCLE

Adult worms are 2–5 cm long, the thinner anterior half of the body being normally partly buried in the mucosa of the large bowel of the host (caecum, colon, rectum). They feed on tissue juices, not blood. Female worms produce several thousands of characteristic eggs per day, which escape in the stools. After about 2 weeks' development in warm moist soil, the eggs are embryonated and infective. The larvae are released from the embryonated eggs after ingestion and take up their definitive habitat without a stage of tissue migration. The time from the ingestion of the eggs to the development of the mature worms is about 3 months. Infection requires the ingestion of soil-ripened eggs and so is commonest in children, who swallow most soil.

CLINICAL EFFECTS

Most infected subjects have no symptoms at all. In heavily infected patients (chiefly children and the mentally subnormal), the main effects are:
1 Diarrhoea with blood but without fever.
2 Rectal prolapse.
3 Anaemia.
4 Stunting of growth.
5 Poor cognitive function.
6 Eosinophilia.

DIAGNOSIS

Direct diagnosis

Because the females are so prolific, even light infections can be detected by finding the typical eggs in the direct faecal smear. There is no need for concentration methods, for if eggs cannot be found on direct examination, the infection has no clinical significance.

In symptomatic infection there are very numerous eggs in the stool. If rectal prolapse is due to trichuriasis, many worms may be seen adhering to the rectal mucosa. In other symptomatic infections, proctoscopy reveals the same situation, as worms only occur in the lower bowel in very heavy infections.

Indirect diagnosis

Even if there were an immunodiagnostic method—and there is not—it would be of no clinical use.

Circumstantial diagnosis

The combination of bloody diarrhoea, an iron-deficiency anaemia and eosinophilia in a small child should raise the possibility of trichuriasis. There are no other helpful features. The same picture can be produced by heavy *Schistosoma mansoni* infections.

TREATMENT

It is not worth treating light infections. For symptomatic disease, and those in whom the parasite is suspected of causing symptoms, the treatment of choice is as follows.

Mebendazole

Dose 100 mg twice daily for 3 days. This dosage is given for all those over the age of 2 years, except for pregnant women in whom it is contraindicated, because of teratogenic effects in rats.

Albendazole

400 mg as a single dose (for adults).

EPIDEMIOLOGY AND CONTROL

These are exactly as for *Ascaris lumbricoides* infection. Only in urban slums does the infection achieve public health importance. There are many much better reasons for improving the state of environmental health.

FURTHER READING

Cooper, E.S., Bundy, D.A.P., Macdonald, T.T., Golden, M.H.N. (1990) Growth suppression in the *Trichuris* dysentery syndrome. *Eur J Clin Nutr* **44**, 138–47.

CHAPTER 20

Strongyloides stercoralis

Strongyloides stercoralis is a worm inhabiting the small bowel of humans in large areas of the tropics and subtropics and it can occur in mines in temperate climates. Infection has been recorded in a British girl who acquired it from running barefoot in an English park in summer.

Almost all infections cause trivial symptoms or none at all. The infection only becomes important in patients immunosuppressed by reason of intercurrent disease, malnutrition or medication.

USUAL LIFE CYCLE

Adults live in the small intestine of humans only. The females, 2 mm long and very slender, live in the mucosa. They lay eggs which soon release microscopic larvae which usually escape at the non-infective (rhabditiform) stage in the faeces. Adult male worms are rapidly expelled and reproduction is probably usually parthenogenetic.

In the hospitable environment of warm, moist soil the larvae develop into free-living male and female worms within a week. The free-living females produce another generation of rhabditiform larvae, which develop into infective filariform larvae under certain environmental conditions. Humans are infected by penetration of the intact skin. Larvae may persist in the soil for many weeks, and the free-living cycle may be repeated many times. *S. stercoralis* is the only common soil-transmitted helminth infecting humans in which the worms can multiply in the free-living stage.

After penetrating the skin, the larvae are carried to the lungs, migrate through the alveoli to reach the bronchial tree, and are swallowed to reach their normal habitat. From initial infection to maturity probably takes less than 4 weeks.

AUTOINFECTION CYCLE

Sometimes the rhabditiform larvae, after their release into the bowel lumen, change into the infective filariform stage. They may then reinfect the same host by either penetrating the perianal skin or the bowel wall.

They then migrate through the tissues and the lungs and re-establish themselves in the intestine as new adult worms. This is how infection can persist for more than 40 years, even in the absence of external reinfection, such as in about 1 in 5 of the ex-prisoners of war of the Japanese who worked on the infamous Thai–Burma railway.

CLINICAL FEATURES

Early infection

1 An itchy eruption at the site of larval penetration. The patient seldom recollects this.

2 Cough and wheeze due to larvae in the lungs. This is also not common. Sometimes it seems that adult females produce eggs in the lungs, and large numbers of larvae may then appear in the sputum.

3 Upper abdominal pain and diarrhoea. A fairly acute steatorrhoea associated with early *Strongyloides* infection is well documented. A strong clue to its aetiology is the presence of a high eosinophilia.

4 Weight loss.

The natural history of this syndrome of early infection, if untreated, is for the condition to subside spontaneously in a few weeks, during which time one assumes that host tolerance and immunity develop. The infection then becomes chronic, and usually asymptomatic.

Established strongyloidiasis

It is impossible to overemphasize that most infections are asymptomatic. But the following symptoms do sometimes occur:

1 Larva currens. This is a characteristic, virtually pathognomonic skin eruption. It is caused by the migration of larvae through the skin during autoinfection. The eruption typically:

(a) is a serpiginous wheal (a raised wiggly line) surrounded by a flare;

(b) is evanescent (comes and goes in a few hours);

(c) is very itchy;

(d) is confined to the trunk between the neck and the knees;

(e) tends to appear in crops at irregular and unpredictable intervals.

2 Intestinal symptoms. These are usually vague, taking the form of irregular bouts of looseness of the stools. Diarrhoea is not constant, and the patient may only recognize that his or her bowels were abnormal in retrospect, when the infection has been eliminated. Bloody diarrhoea is not a feature of uncomplicated chronic strongyloidiasis.

HYPERINFECTION SYNDROME

This occurs when the body loses its defences against autoinfection and there are enormous numbers of successful filariform larvae reinvading the body. In its most fulminating form it has been reported in patients receiving corticosteroid treatment, in those with multiple diseases and malnutrition (the beggar syndrome), in malignant lymphoma and following renal transplantation. The features are:

1 Diarrhoea. (This may take the form of severe steatorrhoea leading to death in a few weeks if not recognized. If the colon is severely affected, there may be numerous haemorrhagic lesions leading to a dysentery-like picture.)

2 Paralytic ileus with diffuse inflammatory thickening of the gut.

3 Gram-negative septicaemia.

4 Serous effusions and bacterial peritonitis.

5 Pulmonary symptoms (cough, wheeze, dyspnoea, haemoptysis) due to massive migration of larvae through the lungs.

6 Specific organ involvement such as when many larvae in the brain cause encephalitis and pyogenic meningitis.

At autopsy, filariform larvae may be found in all the tissues of the body.

In a recent fatal case DRB counted the larvae in the peritoneal exudate: there were approximately 20 000/ml.

In the hyperinfection syndrome there is never eosinophilia, and as the eosinophils are the cells mainly concerned with killing the migrating larvae, this is scarcely surprising. Hyperinfection is largely the result of the eosinopenia.

Treatment with thiabendazole in the hyperinvasion syndrome is often unsuccessful. It is too early to say whether albendazole will be more effective or not.

DIAGNOSIS

Direct diagnosis

1 Direct stool microscopy. Motile larvae may be found on direct saline-smear examination. They must be distinguished from the larvae hatched from hookworm eggs if the specimen is old. Rhabditiform larvae are most common. A patient with diarrhoea may pass adult worms.

2 Microscopy after concentration. A good concentration technique such as the formol-ether technique increases the yield, but as the larvae are killed they are more difficult to recognize at low magnification.

3 Microscopy of duodenal contents. It is more difficult to find the larvae in the stools of patients with symptomatic infections than in those without symptoms. We assume that this is due to most larvae in symptomatic disease reinvading the mucosa in the filariform stage, and so failing to reach the outside in the stools. In such cases, larvae (and even eggs and adult worms) may be recovered by:

(a) duodenal aspiration;

(b) the 'hairy string' test (Enterotest capsules);

(c) jejunal biopsy.

The Enterotest capsule is probably the most effective, and the least unpleasant for the patient. A weighted capsule containing an open-texture nylon string is swallowed, one end being tethered to the cheek. Three hours later the string is withdrawn, the capsule having dissolved and the weight passing on to the outside. If the capsule reached the duodenum, the last 10 cm or more of the string will be stained with bile. The string is then squeezed between thumb and forefinger, and the contents milked on to a slide. (Gloves should be worn.)

This test may reveal *Strongyloides* infection detected in no other way, and is also excellent for diagnosing *Giardia* infection.

4 Stool culture: 5 g of stool is mixed with 5 g of charcoal and a little water, and placed on a filter paper in a 5–6 cm Petri dish. This is placed in a 10 cm dish with a small amount of water surrounding the inner dish. Any larvae present in the specimen will develop into adult worms in a week at room temperature (18–20°C) and second-generation larvae will begin to appear in the outer well in 10–14 days, after swarming over the rim of the inner dish and becoming trapped. Larvae are produced in 6 days at 26°C. This is the most sensitive method for detecting small numbers of larvae in faeces. All the same, the culture may be negative even when larvae are present in the duodenal contents. The Harada–Mori method can also be used, but is less sensitive as it uses less material (see Chapter 18, p. 189).

Indirect diagnosis

Various immunodiagnostic tests have been described, one of the most promising being an ELISA test using antigen prepared from *Strongyloides cebus* larvae. It is very sensitive, but its specificity in areas endemic for other helminths has not yet been thoroughly investigated.

Circumstantial diagnosis

Strongyloidiasis is one of the commoner causes of eosinophilia without obvious cause. But the absence of eosinophilia in the most severe cases greatly reduces the value of the eosinophil count. There are no other investigations that really help. Varying degrees of malabsorption of fats and xylose are sometimes found in patients with definite bowel symptoms.

TREATMENT

Thiabendazole (Mintezol) is effective, but causes troublesome side-effects.

Dose

25 mg/kg twice a day for 3 days. It is available as chewable tablets 500 mg and suspension containing 500 mg in 5 ml. Toxic effects include nausea, vomiting and dizziness.

Individual tolerance varies greatly. There are good results (about 80% cure) with normal infections. The success rate with the hyperinfection syndrome is disappointing, even with prolonged courses.

Albendazole (Zentel) is as effective as thiabendazole, and (at a dose of 400 mg twice daily for 3 days) virtually free of side-effects. Neither drug is safe in pregnancy.

EPIDEMIOLOGY AND CONTROL

Strongyloidiasis is most prevalent in the humid tropics. The free-living cycle does best in wetter conditions than those preferred by hookworms. As most infected people develop immunity to hyper-infection, the intensity of infection does not show progressive build-up on continued exposure. Its major importance is the danger of hyper-infection in asymptomatic patients treated with steroids. It resembles amoebiasis and TB in this.

FURTHER READING

Archibald, L.H. *et al.* (1993) Albendazole is affective treatment for chronic strongyloidiasis. *Q J Med* **86**, 191–5.

Grove, D.I. (ed.) (1989) *Strongyloidiasis: a Major Roundworm Infection.* Taylor and Francis, London, New York, Philadelphia.

Toxocara canis (*and cati*)

Toxocariasis is a cosmopolitan infection of dogs and cats with a round-worm resembling *Ascaris*.

Infection in dogs is mainly in puppies less than 6 months old and in pregnant bitches. Infection is acquired by ingesting embryonated eggs and can cause infection in humans leading to visceral larva migrans (VLM). Transplacental infection of puppies occurs. Conceivably, trans-placental infection in humans could occur also. Humans become infected by ingesting embryonated eggs from soil or from dogs' fur. It is mainly a disease of children. *Toxocara* infection is one of the commonest causes of the VLM syndrome. Human infection from cats is probably of little importance.

VISCERAL LARVA MIGRANS

The larva from the ingested egg is released in the intestine and then goes on a prolonged safari through the tissues, lasting 1–2 years. Worms seldom develop beyond the larval stage, so do not reach maturity in the intestine. There are two distinct syndromes: generalized and ocular.

Generalized VLM

This occurs when there are large numbers of larvae in the body, such as in children with pica (dirt-eating). The disease is self-limiting and the eyes are almost never involved. The symptoms are as follows:
1 Fever.
2 Eosinophilia.
3 Hepatomegaly.
4 Occasionally, asthma.

VLM eye lesions

Larvae trapped in the retina may excite granuloma formation. This may impair vision, and is sometimes mistaken for a retinal tumour. This condition rarely develops in patients with generalized VLM, and the two manifestations of the disease seem to be largely mutually exclusive. The predilection for the retina is not understood. Patients with eye involvement seldom have eosinophilia or other evidence of generalized VLM.

Other ill-effects of *Toxocara* infection

The evidence that the infection causes epilepsy and other organ-specific disorders is unconvincing. So is the claimed association with poliomyelitis.

DIAGNOSIS

Direct diagnosis

Unless material is obtained by liver biopsy, or an eye is enucleated, there is no way of establishing the diagnosis directly. The larvae are confined to the tissues, and their absence from the gut precludes the release of eggs in the faeces.

Indirect diagnosis

Although immunodiagnostic tests of various kinds can detect antibodies in many patients with generalized VLM, the success rate with clinically suspected ocular toxocariasis is very low. This has resulted in a continued state of uncertainty about the importance of *Toxocara* in causing eye disease. High titres of antibody in the vitreous and aqueous are virtually diagnostic.

Circumstantial diagnosis

The only really helpful finding is eosinophilia in generalized VLM.

TREATMENT

Prolonged treatment with thiabendazole in generalized VLM may be folowed by reduction in fever, eosinophilia and liver size. However, relapse often occurs on stopping treatment. Because it is much better tolerated than thiabendazole, albendazole is probably nowadays the drug of first choice.

In ocular *Toxocara* infection, treatment is directed towards reducing the inflammatory response to the putative larva by systemic corticosteroid administration, combined with albendazole treatment. A visible larva can be attacked directly by laser.

EPIDEMIOLOGY AND CONTROL

Because there is no direct diagnostic test, the prevalence and importance of the infection remain uncertain. There seems no doubt that generalized VLM is rare in Europe, and as it leaves no lasting effects, it is doubtful if it justifies a special public health effort to control it. In a large (2.6 million) health authority region in the UK, only 1 case of generalized VLM has been recorded in a 5-year period. The incidence in tropical countries is entirely unknown, although the high prevalence of *Toxocara* in the animal population is beyond doubt.

The size of the problem of ocular *Toxocara* infection in humans remains conjectural because there is no satisfactory diagnostic test for suspected cases, but the finding of antibody in the aqueous may greatly improve the situation in the future. Most ophthalmologists are agreed that blindness caused by *Toxoplasma* infection is of far greater importance than ocular toxocariasis.

Control of toxocariasis must rely on reducing the degree of contact between dog faeces and children. This would entail:

1 Health education.
2 Periodic worming of dogs, especially puppies.
3 Control of stray dogs.
4 Limiting access of dogs to public places.

How much human disease could be prevented by such measures, compared with the distress caused to dog lovers, is at present incalculable.

QUESTIONS, PROBLEMS AND CASES

1 An 18-month-old child in the tropics presents with bloody diarrhoea, without fever or malaise. What causes do you think of, and how do you establish the diagnosis?

2 A 5-year-old girl presents at your clinic with a 6-week history of fever. On examination you find her liver is enlarged to 5 cm below the left costal margin, her temperature is 38.6°C, but she seems otherwise well. Routine investigations show a negative blood film, but she has a total WBC of 38 000/μl and the differential count shows 90% eosinophils. What questions would you ask her mother? What other investigations would you do?

3 (a) What parasite(s) can cause uniocular blindness in children? (b) How would you distinguish them by the use of special tests?

4 A child in the tropics vomits up a roundworm which is 14 cm long. What is it likely to be?

5 A 5-year-old child presents with an enlarged, tender liver and fever. List the main diagnoses possible, together with the important diagnostic findings.

6 A 20-year-old farmer in the tropics presents to you with a 3-week history of diarrhoea and upper abdominal pain. Your routine tests show that he has an eosinophilia of 65% of a total WBC of 9000/µl.
(a) What causes do you suspect?
(b) How do you establish the diagnosis?

7 A 42-year-old African woman with a long history of severe asthma is admitted to your hospital in status asthmaticus. Five days after starting high-dose prednisolone phosphate treatment, she develops severe diarrhoea and begins to go downhill.
(a) What parasitic infection could be the cause of this, and how would you establish the diagnosis?
(b) Suppose her diarrhoea were bloody, what would you suspect and how would you confirm it?

8 What are the likely diagnoses of an itchy, raised and erythematous rash which keeps recurring for some weeks in a patient recently returned to Europe after a 3-month scientific survey in a game park in Tanzania?

9 What are the dangers of oral iron overdose in young children? Name the best antidote.

10 You have treated a patient with HWA with iron and an appropriate anthelmintic, but the haemoglobin level has only risen from 7 to 10 g/dl. It remains unchanged after further well-supervised iron treatment. What causes would you suspect to explain this failure of treatment?

11 An obviously anaemic farmer attends your clinic. What simple laboratory investigations would lead you to a diagnosis of HWA?

Tapeworms

Taenia saginata: *the beef tapeworm*

Taenia saginata is a cosmopolitan infection in which humans harbour the adult worm and cattle harbour the larval stage. Its main importance is in economic losses caused by condemnation of beef carcasses. Human infection is of social importance only. Ethiopia has the highest infection rate in the world.

PARASITE AND LIFE CYCLE

People acquire infection by eating undercooked meat containing cysticerci, the larval stages of the parasite encysted in the muscles of infected herbivores. The cysts evaginate in the intestine, and the head of the worm attaches itself to the mucosa of the upper third of the small intestine by its suckers. Segments called proglottides grow from the head, and new segments are added until the worm contains a chain of 1000–2000. A full-grown worm is often more than 5 m long, sometimes 10 m.

Proglottides at the tail end of the worm develop fertilized eggs in the uterus and are called gravid segments. When mature, the gravid segments break off the chain (strobila) and leave the anus in the stool or by their own movements. Sometimes proglottides rupture in the intestine, and free eggs are also passed in the stool. The eggs which reach pasture, mainly after disintegration of the mature proglottides, are infective to cattle (and several other herbivores) when swallowed. They hatch in the bovine gut to become oncospheres, and enter the circulation where they are carried to the muscles and encyst as cysticerci. The condition in the infected animal is cysticercosis from *Cysticercus bovis*.

The meat is described as 'measly', and the cysticerci are easily visible to the naked eye.

CLINICAL PICTURE

Most infections come to light because the patient notices the motile white proglottides in the stool, or suffers embarrassment and dismay when they force their way out of the anus unbid. The worms do not suck blood, and rarely cause serious pathology in the gut. Various vague abdominal symptoms such as 'hunger pains' have been blamed on the worm, and in rare cases they may cause intestinal obstruction. The worm becomes sufficiently mature to shed segments about 12 weeks after infection.

DIAGNOSIS

The eggs found in the stool are indistinguishable from those of *Taenia solium*, the pork tapeworm. To make a specific diagnosis, a mature proglottid is pressed between two microscope slides and the number of lateral branches of the uterus counted. *T. saginata* has 15–20 main branches on each side; *T. solium* has 13 or fewer, but this criterion is not as reliable as once thought.

TREATMENT

Niclosamide (Yomesan) is the drug of choice. The dose for all ages over 6 years is 2 g, taken before breakfast as two doses of 1 g separated by a 1-h period. The 500 mg tablets should be well chewed, and washed down with a little water. For ages 2–6 years, the dose is 1 g; under 2 years, 500 mg. The drug is not absorbed, and only minor gastrointestinal side-effects occur. The drug kills the worm, which is normally expelled in an unrecognizable, partly digested state. There is no need to purge, unless someone wants an intact worm, in which case, give a saline purge 2 h after the drug.

Praziquantel or albendazole may also be used.

EPIDEMIOLOGY AND CONTROL

The disease occurs where people eat raw or very undercooked meat from cattle, perhaps occasionally from reindeer and other herbivores. Control measures are obviously environmental sanitation and meat inspection. Immediately after niclosamide treatment, the stool contains countless thousands of eggs, and special care should be taken to ensure that it is safely disposed of.

There is evidence that seagulls feeding on sewage outflows containing proglottides may help to disseminate the infection on to pastures in the UK.

CHAPTER 23

Taenia solium:
the pork tapeworm

Taenia solium is a much less common infection than *T. saginata* but far more important because of its ability to cause severe disease in humans. Humans are the definitive host, the pig the normal intermediate host. It is found all over the world where people eat raw or undercooked pork. For this reason, it is rare in Muslims and Orthodox Jews; although some Orthodox Jews in New York were infected by housemaids who harboured adult tapeworms in their gut. In some parts of the world such as central America cysticercosis from *T. solium* is a major cause of epilepsy and other neurological disease.

PARASITE AND LIFE CYCLE

The worm resembles *T. saginata*, with relatively minor differences in morphology such as the uterine structure mentioned, and it tends to be smaller. The life cycle, substituting the pig for cattle, is also similar to that of *T. saginata*, with one great difference: humans can become infected by the larval stage.

If someone swallows eggs of *T. saginata*, they do not develop into cysticerci. If someone swallows the eggs of *T. solium*, they do develop into cysticerci, causing the disease cysticerosis. The larval (bladder worm) stage of the parasite is called *Cysticercus cellulosae*. Human cysticercosis can develop in two ways:

1 From swallowing other people's *T. solium* eggs. This may be from contaminated food, but infection during oral–anal contact in homo- and heterosexual relations is increasingly recognized as common.

2 From swallowing one's own *T. solium* eggs in food contaminated by faecally fouled fingers. Internal autoinfection, occurring entirely in the gut, is a hypothetical possibility which has never been substantiated.

Infection with the adult worm

Symptoms are usually trivial, unless cysticercosis develops. They resemble those in *T. saginata*. The passage of proglottides usually brings the patient to the doctor.

Human cysticercosis

There are commonly no symptoms at all in light infections. The bladder worms encyst in muscles and subcutaneous tissues and there is little inflammatory response. The first indication of infection is commonly the finding of small (0.5–1 cm) swellings beneath the skin, or their chance discovery by X-rays some years later, when they have calcified in skeletal muscles.

Cerebral cysticercosis usually presents as epilepsy, of any type, 3 or more years after initial infection. Ill-defined focal neurological symptoms (and less often, physical signs) may also develop. It may be very difficult to decide how much of the symptomatology is organic, and how much is due to anxiety. Fits usually continue indefinitely, but sometimes cease spontaneously. Rarely, cysticerci in other sites such as the eye and heart cause relevant symptoms.

DIAGNOSIS

Adult worm

Finding typical *Taenia* eggs in the stool makes the diagnosis of taeniasis. Identifying the proglottides (see Chapter 22, p. 210) incriminates *T. solium*.

Cysticercosis

Direct diagnosis is not possible unless cysticerci in nodules can be excised and identified histologically. Various immunodiagnostic tests may be positive. Immunoelectrotransfer blot seems to give good results on serum or CSF.

Diagnosis is usually made from circumstantial evidence, especially the finding of typical spindle-shaped calcifications in muscles, most numerous in the thighs. Calcification in muscles usually appears 3–5 years after initial infection. Calcified cysticerci are less often seen in the brain: in about one-third of cases, 10 years or more after infection. Computed tomography (CT) scans are very useful in diagnosis when available.

TREATMENT

Adult worm

It used to be thought that treatment with a drug that kills the worm and leads to its destruction in the bowel lumen, such as niclosamide, would put the patient at risk of developing cysticercosis from the liberation of the eggs. Such fears have proved unfounded, perhaps because the eggs would have to be digested in the stomach before releasing the oncospheres. All the same, it is still wise to anticipate this possibility by using the following regimen:

1 Give an effective antiemetic on waking.

2 1 h later: give niclosamide exactly as described for *T. saginata* (see Chapter 22, p. 210).

3 2 h after the second dose, give a saline purge in full dosage.

The worm is then normally expelled intact, and cysticercosis has not been a problem. Praziquantel or albendazole are alternative treatments.

TEST OF CURE AND FOLLOW-UP

It takes about 3 months for the worm to regrow if the strobila is expelled but the living scolex is left behind. This used to be a problem with the older remedies, following which the stool had to be diligently searched to ensure that the scolex had been expelled. With niclosamide, which kills rather than paralyses the worm, the problem does not arise.

Cysticercosis

Until recently the only treatment for cysticercosis was symptomatic — anticonvulsant therapy to control fits. Although it has been known for several years that praziquantel can destroy cysticerci in animals, it was feared that the drug might produce fatal reactions in cerebral cysticercosis. Numerous studies have now shown that, when used in people, praziquantel can lead to remission of symptoms and (radiologically proven by CT scans) regression of the cysts. But during treatment and for a week or two after, there may be severe headache and aggravation of symptoms. Clearly, the reaction around the dying cysts could lead to dramatic complications such as acute obstructive hydrocephalus. For this reason, praziquantel must only be given under careful supervision, preferably where neurosurgical help is available.

The dose used has been 50 mg/kg daily, in three divided doses, for 21 days. Albendazole, as a dose of 15 mg/kg daily for 30 days, is perhaps even more effective. If severe headache develops dexamethasone is started immediately.

CONTROL

Control relies on:

1 Health education.
2 Environmental sanitation.
3 Effective meat inspection.
4 Effective treatment of patients harbouring adult worms.

FURTHER READING

Garcia, H.H., Martinez, M., Gilman, R. *et al.* (1991) Diagnosis of cysticercosis in endemic regions. *Lancet* **338**, 549–51.
Vaquez, V., Sotelo, J. (1992) The course of seizures after treatment for cerebral cysticercosis. *N Engl J Med* **327**, 696–701.

CHAPTER 24

Hydatid disease: Echinococcus infections

Hydatid disease is due to the larval stage of a small tapeworm of dogs and other canines developing in humans. The infection is a zoonosis, normally maintained in dogs and sheep or cattle in close association with humans (*Echinococcus granulosus*), or in a wild cycle such as in wild canines and rodents (*E. multilocularis*). Most human infections are with *E. granulosus* and are associated with the rearing of sheep and cattle in climatic conditions varying from tropical to subarctic.

LIFE CYCLE

E. granulosus

Infected dogs harbour the minute adult tapeworms, 3–6 mm long, in their small intestine. The worms possess only three proglottides, the end one being mature. The eggs are liberated either before or after the proglottid escapes in the faeces, and contaminate pasture. When ingested by the normal herbivorous intermediate host, the oncospheres liberated in the gut enter the circulation and are trapped in the capillaries of various viscera, where they develop into cysts. A cyst is composed of a sphere of germinal epithelium containing protruding invaginations (brood capsules) and fluid. From the inner surface of the brood capsules, protoscolices develop, invaginated in much the same way as the cysticerci of *Taenia* spp. The whole structure is a hydatid cyst, and it becomes surrounded by fibrous capsule derived from the host tissue. The cyst may develop large daughter cysts in its cavity, each containing more brood capsules. The cyst continues growing for years. Brood capsules which break free from the cyst wall, and individual scolices in the cyst cavity, are called hydatid sand.

Dogs become infected by eating the contents of hydatid cysts in infected carcasses. Sheep or other herbivores become infected by swallowing the taenia-like eggs passed in dog faeces. The strain of *E. granulosus* in the UK which commonly infects horses and has the foxhound as its definitive host is probably not pathogenic for humans. There are several other biological complexes in nature which probably do not pose the risk of human infection.

E. multilocularis

This is similar, but different hosts are involved. The cyst produces daughter cysts by external and not internal budding, so it tends to extend like a malignant tumour, progressively invading surrounding tissues. It is not contained in a well-defined, fibrous capsule. Humans become infected by swallowing the eggs passed by foxes and other *Canidae*, possibly mainly from contaminated wild ground fruits such as bilberries and their close relatives, the lingonberries and cloudberries, widely eaten in northern Europe. An intense focus of infection has been described in Gansu province, China. Various rodents are the intermediate hosts.

CLINICAL PICTURE

About 70% of cysts develop in the liver, 20% in the lungs, and the rest in rarer sites. The clinical features depend on three processes:

1 Mechanical effects such as painful enlargement of the liver (the commonest presentation), cough and breathlessness (cysts may compress bronchi and cause collapse of a lobe), symptoms suggesting brain tumour (intracerebral cysts), bone pain or spontaneous fractures (medullary cysts of bone).

2 Allergic processes due to escape of allergenic hydatid fluid into the circulation, such as urticaria and anaphylaxis.

3 Complication due to:
(a) rupture causing anaphylactic shock and sudden death;
(b) rupture causing spread to other organs or serous cavities due to seeding with viable germinal epithelium;
(c) secondary infection of the cyst.

In endemic areas, hydatid cysts are often much the commonest cause of intra-abdominal masses and solid-looking pulmonary shadows. Some cysts will continue to grow until they rupture and lead to the patient's death. Some will stop growing, the germinal epithelium will die, and the gradual involution of the cyst leaves only a harmless calcified remnant behind. The fate of an individual cyst is unpredictable, and some cysts seem to be sterile from the start.

DIAGNOSIS

Direct diagnosis

This is only possible when the cyst is removed at operation. Aspiration of the cyst contents should not be carried out for diagnostic purposes, because subsequent leakage may cause anaphylaxis or the development of widespread metastatic cysts.

Indirect diagnosis

1 The Casoni test: this involves injecting 0.1 ml of standardized Seitz-filtered hydatid fluid intradermally. A positive result is the development of a wheal, at least twice the diameter of the initial bleb and usually surrounded by a pronounced flare, within 20 min of the injection. The test remains positive long after all cysts have been removed and is unreliable.

2 The detection of antihydatid antibodies: several methods are in use. At present countercurrent immunoelectrophoresis (CIE), the complement fixation test (CFT) and ELISA test are most in favour. Antibody tests are best for liver hydatid but false-negative results occur. The CIE method has unfortunately given false-positive results with lung cancer – an important clinical differential diagnosis – even with the most specific arc 5.

3 The detection of circulating antigen. This may be positive even when antibody tests are negative.

Circumstantial diagnosis

There are no really useful tests, apart from the presence of suggestive radiological findings and eosinophilia. Eosinophilia is not present in all cases. Ultrasound or CT scanning may reveal the presence of intra-abdominal cysts undetectable by other means. Typical findings of daughter cysts within the large cyst make hydatid disease extremely probable, but the nature of the cyst is not always apparent.

TREATMENT

The drug albendazole shows great promise, leading to regression of the cyst in many cases. A regimen of 400 mg twice daily (or 10 mg/kg per day) for 4 weeks is commonly used repeated to a total of three cycles (initially) with 2-week breaks in between. Liver function and WBC should be monitored throughout. Where possible a cycle of albendazole treatment should be given before surgery to minimize the risk of recurrence.

Accessible cysts are excised intact. The cyst should be sterilized chemically once it has been exposed but, since many cysts connect to the biliary system, chemical sterilization must not be used if the contents of the cyst are bile-stained or purulent. Cetrimide 0.1% or silver nitrate 0.5% is used to replace the aspirated cyst fluid as scolicides. If the cyst is ruptured during surgery, death from anaphylaxis or the delayed results of dissemination may follow. Where the surgeon is competent, multiple cysts confined to one lobe of the liver are best dealt with by hemihepatectomy.

EPIDEMIOLOGY AND CONTROL

As all the factors causing human infection are well known, it might seem surprising that the disease is still endemic in the UK. It persists in areas of hill sheep-farming, where sheepdogs are infected by feeding on cyst-containing carcasses of sheep which die on the hillsides. In the past, in many countries, infection was maintained by deliberately feeding infected carcasses to dogs. This practice has been prevented by intensive health education in areas previously hyperendemic, such as Iceland and New Zealand. Very high levels of human infection persist in areas where sheep- and cattle-rearing and the presence of domestic dogs coexist in the absence of effective health education, such as in the Turkana region of Northern Kenya. There, the very close association of dogs and children ensures that children are exposed to the eggs, and infections develop at an early age.

The third main control measure is the periodic worming of dogs. The importance of the fox in maintaining *E. granulosus* is probably small.

FURTHER READING

Gil-Grande, L.A., Rodriguez-Caabiero, J., Prieto, J.G. *et al.* (1993) Randomised controlled trial of efficacy of albendazole in intra-abdominal hydatid disease. *Lancet* **342**, 1269–72.

Morris, D.L., Richards, K.S. (1992) *Hydatid Disease, Current Medical and Surgical Management.* Butterworth-Heinemann, Oxford.

SECTION 5

Flukes

CHAPTER 25

Blood flukes: schistosomes

HUMAN SCHISTOSOMIASIS (BILHARZIA, BILHARZIASIS)

Adult flukes are white, worm-like creatures 1–2 cm long which inhabit parts of the venous system of humans. The male worm resembles a rolled leaf in having a groove on his ventral surface in which the longer, more slender female is held *in copulo*. Both sexes are actively motile.

The worms sometimes live for 30 years, but their normal life span is probably 3–5 years. Three main species affect humans:

1 *Schistosoma haematobium*, which causes urinary schistosomiasis. It is scattered throughout Africa, parts of Arabia, the Near East, Madagascar and Mauritius.

2 *S. mansoni*, which is mainly found in Africa and Madagascar. It was exported by the slave trade to parts of south America, the Caribbean and Arabia.

3 *S. japonicum*, which is found in China, Japan, the Philippines and Sulawesi. There is a small focus in the Mekong river on the east border of Thailand. *S. mansoni* and *S. japonicum* cause disease of the bowel and liver.

(*S. intercalatum* is a minor species confined to west Africa. It inhabits the veins of the lower bowel and produces terminal-spined eggs.)

Two hundred million to three hundred million people are infected. The disease is spreading, as improved water supplies create new habitats for snails.

LIFE CYCLE

Fertilized adult females lay eggs in the terminal venules of the preferred host tissues. Their bodies obstruct the vessel and so impede the escape of eggs into the circulation. Most of the eggs penetrate the vessel wall and enter the tissues. Movements of the walls of the hollow viscus involved propel the eggs towards the lumen, from which they escape to the outside world—in the urine in the case of *S. haematobium*, in the stools in the case of *S. mansoni* and *S. japonicum*.

The shapes of the eggs of each species are distinctive, but each contains a virtually identical ciliated miracidium. This hatches out in fresh water, and swims in search of a suitable snail intermediate host. Many species of snail host are known, but in general:

1 S. *haematobium* requires an aquatic, sinistral, turretted snail of the genus *Bulinus*.

2 S. *mansoni* requires a flat, aquatic, 'ramshorn' snail, most commonly of the genus *Biomphalaria*.

3 S. *japonicum* requires a small, amphibious, operculate, turretted snail, usually of the genus *Oncomelania*.

The miracidium penetrates the body of the snail and begins a complicated, asexual, replicative cycle which results, a few weeks later, in the release of minute, fork-tailed cercariae into the water. As cercariae are about 200 μm long, they are just visible to a young eye. They emerge from the sporocyst inside the snail in response to light. A snail may shed cercariae for many weeks. The cercaria is infective to the definitive host. If it finds no suitable host within 24–48 h, it dies.

If it contacts human skin the cercaria penetrates, using enzymes contained in special glands, sheds its tail and body, and enters the circulation with the new name—schistosomule.

The schistosomule reaches the liver through the lungs, by a disputed route. Once in the liver, it begins to feed and grow, and in 1–3 months develops into a mature fluke in an intrahepatic portal vein. The mature males and females couple, and then migrate to their final habitats.

It is easy to see how S. *mansoni* and S. *japonicum* find the way to their homes in the lower mesenteric veins, as they have only to travel straight down the portal vein. It is a mystery how S. *haematobium* reaches the vesical plexus.

The time elapsing between cercarial penetration and the passage of eggs is the prepatent period. It can be as short as 4 weeks with S. *mansoni* (Fig. 25.1), usually 12 or more weeks with S. *haematobium*, and somewhere in between with S. *japonicum*. The prepatent period is sometimes very prolonged in light infections; perhaps if there are few worms in the liver, the sexes have difficulty finding each other.

EFFECTS OF PENETRATION

Cercariae may cause an itchy papular rash—'swimmer's itch' or 'fisherman's itch'. This is seldom seen in endemic areas. A conspicuous cercarial rash is more often due to avian or other schistosomes not otherwise pathogenic in humans. It is quite common in north Europe, north America and South-east Asia.

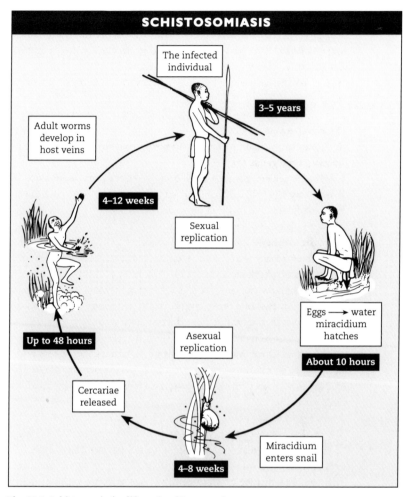

Fig. 25.1 Schistosomiasis – life cycle of S. *mansoni*.

Initial illness: immune complex disease?

An initial illness occurs most often in S. *mansoni* and S. *japonicum* infection, and is usually only recognized following the first exposure. It does not develop in very light infections, and is seldom recognized in residents of endemic areas. It is mainly a problem in immigrants or visitors encountering a large cercarial challenge for the first time.

The illness comes on 4 or more weeks after infection, and is usually self-limiting. The theory is that, as the worms begin to lay eggs, soluble antigen leaks out of the eggs and into the circulation. While antibody production lags behind antigen release, moderate antigen excess prevails. This favours Ag/Ab complex formation, with the development of generalized immune-complex disease. Because the antigen is soluble and distributed by the blood stream, the effects are more general than local.

Features of initial illness

The condition is sometimes called Katayama fever, after the prefecture in Hiroshima ken in Japan where it used to be common. Some or all of the following may occur:

1 Fever.
2 Urticaria.
3 Eosinophilia.
4 Diarrhoea.
5 Hepatomegaly.
6 Splenomegaly.
7 Cough and wheeze.
8 Cachexia.

Perhaps it is seldom recognized in children in endemic areas because immune tolerance develops *in utero*, due to transplacental passage of antigen.

Spontaneous recovery may be related to restoration of Ag/Ab balance as the infection matures and antibody production increases. This ensures that immune complex is precipitated around the eggs, and not more generally.

Importance of the eggs: those that get away

Eggs which escape from the body enable the life cycle to be completed. Their passage through the bladder in S. *haematobium* typically causes terminal haematuria, the cardinal symptom of the infection. In heavy infections, irritation of the bladder may cause dysuria.

Rarely, with a very low dietary iron intake, iron-deficiency anaemia may result.

In S. *mansoni* and S. *japonicum* corresponding effects may occur in the bowel: diarrhoea and blood. More commonly, the presence of a little blood is noticed in an otherwise normal stool. In most infections, no bowel symptoms are noticed.

Importance of retained eggs: the main pathology

The serious mischief done by schistosomiasis is from tissue reaction to retained eggs. This reaction, which follows sensitization to egg antigens, is a circumoval granuloma. It results from combined humoral and CMI

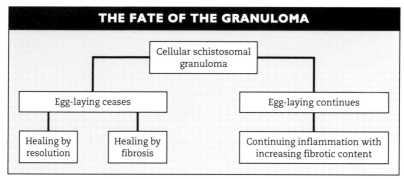

Fig. 25.2 The fate of the granuloma.

attack on the egg, and the granuloma occupies several hundred times the volume of the egg itself. Its characteristics are:

Epithelioid and giant cells ⎫
Lymphocytes   ⎬ arranged in concentric fashion
Eosinophils   ⎭ around the egg.

The cellular content diminishes with time, to be replaced by fibroblasts and a collagenous scar. Precipitation of Ag/Ab complex on the egg surface helps activate inflammation.

The duration of the vigorous cellular response to a single egg lasts a few weeks. If egg-laying is stopped by chemotherapy, the cellular component of the granuloma usually resolves in 2 or 3 months. Not all granulomata lead to scars. Fig. 25.2 shows what happens.

These pathological processes occur, with variations, in all the schistosome infections. They help to explain the specific features described next.

SPECIFIC FEATURES

S. haematobium: obstructive uropathy

Eggs are laid in the bladder and nearby organs, not singly, but usually in clutches. This is because a female schistosome may occupy the same site for long periods, during which time she lays several hundred eggs a day. The eggs give rise to a granulomatous lesion up to several centimetres in diameter. Most commonly these fleshy lesions form in the bladder mucosa, where they simulate tumours and so are called pseudo-papillomata. They may be sessile (flat) or pedunculated (on stalks). Smaller deposits of eggs cause lesions a few millimetres in diameter, resembling tubercles.

When granulomata form near the ureteric orifices or in the ureters themselves, the ureters may become obstructed. This is the cause of early obstructive uropathy.

The secondary effects are hydroureter, in which the ureter becomes dilated and elongated, and varying degrees of hydronephrosis. In the most severe cases, kidney drainage may be so impaired as to cause uraemia.

NATURAL HISTORY OF OBSTRUCTIVE UROPATHY

It used to be thought that all the changes of obstructive uropathy were irreversible. This is because:

1 Similar obstructive lesions in Europe and north America are usually irreversible.

2 In schistosomiasis-endemic areas, obstructive uropathy is radio-logically more common in children than in adults. The natural inference is that most children with obstruction die before adulthood.

It is now known that in the early cellular phase of the granuloma, complete resolution may follow effective chemotherapy. Longitudinal follow-up has shown that, provided reinfection does not occur, spontaneous resolution without significant scarring may also occur. This must be related to the usually short (3 years) life span of the worms.

FACTORS AFFECTING REVERSIBILITY

The main factor is the duration of infection. Intensity is related more to the incidence and severity of obstructive lesions than to reversibility. The duration of infection is mainly related to the pattern of exposure. The evidence that different geographic strains of schistosomes vary in their pathogenicity is inconclusive.

DIMINISHING EXPOSURE: THE CHILDHOOD FACTOR

In hot countries, children naturally play in water. This recreational exposure is sometimes the most important source of infection.

At puberty, exposure often diminishes as modesty develops at the same time as sexual awareness. Infection may still be acquired, however, during activities such as personal bathing, washing clothes and utensils, and in the pursuit of irrigated farming. This change in behaviour is one factor in the tendency of the infection to diminish after puberty. But it is not the only one. Acquired immunity also appears to reduce the likelihood of a given exposure to cercariae leading to an established infection.

Calcification of the bladder

This is common in S. *haematobium* due to calcification of the eggs, not of the bladder itself. A calcified bladder outline on X-ray is fully compatible with normal bladder function. Calcification tends to disappear over the years, provided reinfection ceases or is greatly reduced, presumably due to the continued passage of calcified eggs in the urine (we have often observed this).

The ureters may be calcified in heavy infections, and the calcification may demonstrate the dilatation and tortuosity characteristic of hydroureter.

Late effects of urinary schistosomiasis

These are usually only seen in heavy infections of long duration.
1 Irreversible obstructive uropathy due to fibrous stricture formation, including hydroureter, hydronephrosis and non-functioning kidney.
2 Reduced bladder capacity due to thickening and rigidity of the bladder wall.
3 Carcinoma of the bladder. This is of squamous cell type, and occurs two decades earlier than the transitional cell carcinoma most common in Europe.
4 Kidney stones associated with chronic obstruction. The causal relationship between bladder stones and S. haematobium is unproven.

Metastatic eggs in S. haematobium: schistosomal cor pulmonale

Eggs escaping from the pelvic veins into the caval circulation reach the lungs. In heavy, prolonged infections, granuloma formation may cause obstruction in pulmonary arterioles. Pulmonary hypertension, right ventricular hypertrophy and congestive heart failure may follow. Cyanosis develops from vascular shunting in the lungs.

The chest X-rays show prominent pulmonary arteries, perhaps right ventricular enlargement, and sometimes scattered shadows in the lung parenchyma.

S. MANSONI INFECTION: USUAL PICTURE

Most patients with S. mansoni infections have few or no symptoms. Blood loss may contribute to iron-deficiency anaemia. The liver is often enlarged, the spleen only in the presence of portal hypertension, or during the initial illness. Severe clinical effects, except those due to ectopic worms, are only seen in heavy infections.

PSEUDOPOLYPOSIS OF THE COLON

In severe S. mansoni, granulomata in the large gut may develop into papilloma-like outgrowths of the mucosa. They may ulcerate and bleed, and cause symptoms of dysentery. There is no proven causal relationship to colonic carcinoma, and strictures do not form.

S. MANSONI AND THE LIVER

Severe and long-standing S. mansoni infections cause a characteristic liver disease called Symmers pipestem fibrosis. Large numbers of eggs reaching the periportal regions cause a granulomatous response which leads to gradual occlusion of the intrahepatic portal veins. Portal hypertension follows, but liver cell function is not disturbed.

The clinical features are:

1 Enlargement of liver and spleen.

2 Bleeding from oesophageal varices.

Patients tend to survive their bleeds much better than patients with true cirrhosis, because of the well preserved hepatocellular function. Also, because the serum albumin level is well maintained, ascites is not typical until the terminal stages of the disease.

In late cases, hepatic perfusion may be so impeded that peripheral liver ischaemia occurs. Then, features of true cirrhosis may develop.

When portocaval shunts are well established, the eggs of S. mansoni may bypass the liver in large numbers, and so reach the lungs. In some cases, they may be numerous enough to cause schistosomal cor pulmonale.

Ectopic worms

These have wandered from their usual habitats and taken up residence elsewhere. The chances are increased in heavy infections, where worms may be crowded out.

The most important site for ectopic worms is the CNS, such as in the paravertebral venous plexus or the cerebral cortical veins.

All three species of schistosome may be found in the paravertebral plexus, where egg-laying leads to the development of a granuloma in the constricted space of the spinal canal. The clinical syndrome is spinal cord compression or a cauda equina lesion. If treated promptly, full functional recovery may occur. When it does not, ischaemic injury may be the cause.

S. japonicum is most often ectopic in the brain. A large localized granuloma produces symptoms and signs indistinguishable from cerebral tumour. Diagnosis is seldom made before operation, and the response to specific treatment is usually dramatic.

Tissue digestion studies of schistosomiasis cases at postmortem have shown that virtually all the organs contain schistosome eggs. The relation of eggs found in the brain in this way to symptoms such as epilepsy and neurosis remains speculative.

Other sites for ectopic worms include the subserosal veins of the abdomen, leading to peritoneal granulomata resembling tumours, and the female genital tract where granulomata of the tubes may cause sterility. Cutaneous lesions with an itchy, punctate skin eruption may also occur.

Adult worms trapped in the lungs do not thrive and soon die.

S. JAPONICUM INFECTION

This resembles S. *mansoni*, but with an equal number of worms the infection is more severe. The parasite is less well adapted to humans, the circumoval granuloma is very large, and the egg output of each female worm is greater than that in S. *mansoni*. The initial illness (Katayama fever) may be prolonged and sometimes fatal. Many Chinese workers believe that S. *japonicum* can cause carcinoma of the colon, but most other experts consider the case is unproven.

Salmonella infections and schistosomiasis

Various *Salmonella* spp. can escape the host's immune defences and attack by antibiotics in people harbouring schistosome infections. Perhaps the bacilli lurk in the granulomata, but whatever the reason, it is often impossible to eradicate the *Salmonella* infection until the schistosomiasis has been treated. The commonest associations are:

1 Chronic urinary *Salmonella typhi* carrier state and *Schistosoma haematobium* infection.

2 Chronic *Salmonella* (various species) septicaemia and *Schistosoma mansoni* infection.

The kidneys in schistosomiasis

The following associations are recognized:

1 S. *haematobium*:
   (a) hydronephrosis;
   (b) pyelonephritis;
   (c) nephrotic syndrome (reversible after specific chemotherapy; reported in Somalia).

2 S. *mansoni*:
   (a) nephrotic syndrome (immune-complex deposition) in otherwise uncomplicated infections;
   (b) nephrotic syndrome associated with chronic *Salmonella* infection and focal glomerulonephritis;
   (c) amyloidosis.

It is highly probable that *Schistosoma japonicum* infection will resemble *S. mansoni*, but its renal aspects have received little attention.

Immunity in schistosomiasis

There is evidence that some degree of resistance to superinfection develops in schistosomiasis. It is certainly incomplete, and may depend for its maintenance on the continued presence of some living schistosomes in the body—so-called concomitant immunity. It does not seem to be antibody-mediated, and is probably directed against the schistosomule stage. The survival of adult worms in the circulation may be partly related to their ability to incorporate host antigens in their integument (surface).

In a population apparently exposed to a uniform risk of infection, some people will be found to be very heavily infected and some very lightly infected. If egg output is accepted as being related to the number of adult worms present, the distribution of worms in the population is not 'normal'. It will always be found that some of those infected have an infection with perhaps 100 times as many worms as those with the most common level of infection. If a frequency–distribution plot of egg output is done, the usual Gaussian curve will be seen to be distorted by having a greatly extended 'tail' to the right of the graph. The curve can be made to resemble a 'normal' curve if, instead of the egg count, the logarithm of the egg count is plotted on the x axis. This sort of distribution is called a log-normal distribution, and applies to the abundance of almost all non-replicative parasites in humans and animals.

Mystery of the log-normal distribution

There is no generally accepted explanation, but it could be related to the host's first exposure to infection. If the initial challenge was with a large number of parasites, at a time when no immunity existed, a large population could become established in the absence of immune opposition.

On the other hand, if the first exposure was to a small number of parasites, the subsequent development of immunity could resist further infective challenges, and the total number of parasites would then remain low. The log-normal distribution could then be an expression of the contagious (non-Poisson) distribution of cercariae in water.

Natural history of schistosomiasis

Most people infected with schistosomiasis will die of an unrelated disease. In many parts of the world, although the prevalence is high, the adverse effects of the infection are difficult or impossible to demonstrate. The notion that the infection always causes general debility and malaise, in the absence of more specific effects, is wrong. In the absence of reinfection, the tendency is for most of the worms to die within a few years and for the circumoval granulomata to resolve. In only a small proportion of cases will progressive pathology develop. These are mainly those with heavy infections and re-exposure to infection over a period of many years.

Treatment can certainly modify the natural history of the infection and even advanced cases may show a surprising degree of improvement after chemotherapy.

DIAGNOSIS

Direct diagnosis

This is much the most important diagnosis method. The adult worms are usually inaccessible, so the aim is to find living eggs. The miracidium inside the egg dies within 4 weeks.

URINE FOR S. HAEMATOBIUM

Most eggs are voided around midday. Specimens collected between midday and 2 p.m. are most likely to contain eggs. For quantitative surveys, it is very important to standardize urine collection times. Exercise has no demonstrable effect on egg output. There is no significant or reliable concentration of eggs in any part of the urinary stream.

The methods for finding eggs depend on their high specific gravity (sedimentation) or their size (filtration).

Urine sedimentation

Eggs are sedimented by natural gravity (30 min in a conical glass; the sediment is then aspirated by Pasteur pipette and examined with a ×10 objective under coverslip), or by artificial gravity (10 ml urine in 15 ml centrifuge tube, spun for 3 min at 1500 rpm, arm radius not critical; the deposit is then examined as before).

Living eggs are translucent, and the miracidium is recognizable. Flame cells can be seen flickering.

The viability of the eggs can be checked by adding them to boiled (cool) water in a flask. Emerging miracidia are visible in light shone across the neck. Normal-looking eggs usually hatch. Opaque (calcified) eggs do not, and do not themselves signify active infection (live worms).

Urine filtration

Urine is passed through Whatman 541 filter paper by vacuum or pressure. It is then stained purple by incubation in ninhydrin solution. The entire 24-h urine can be filtered. There are several variants of this method. The WHO has now devised a miniature membrane version of the filtration method, which allows the eggs to be detected unstained.

Advantages

It is very sensitive, accurate for counts and a permanent record is available.

Disadvantages

Cost and time.

STOOL FOR S. MANSONI AND S. JAPONICUM

Direct smear examination is not sufficiently sensitive: e.g. output 100 000 eggs per day; stool 200 g; smear 2 mg. Average count 1 egg per smear. There is a one in three chance of finding no eggs in a case who could be harbouring about 1000 female S. mansoni.

Instead, use a sensitive concentration technique such as formol-ether, TIF (thiomersal, iodine, formol) glycerol sedimentation (the simplest) or a modified Kato smear. Consult a large textbook for details. False positives can occur if the patient has recently eaten liver containing schistosome eggs.

RECTAL BIOPSY FOR ALL SCHISTOSOMES

A small piece of rectal mucosa is removed by biopsy forceps or curette under direct proctoscopic vision. It is placed on a slide under a coverslip or another slide and examined with a ×10 objective.

S. haematobium eggs are often trapped in the rectal mucosa, but may be calcified. It can be difficult to identify living eggs. S. mansoni and S. japonicum more often look viable.

OTHER BIOPSY MATERIAL

Histology is not used for diagnosis as a routine. If a schistosome egg is suspected in biopsy material, take 20 μm sections to maximize the chance of seeing the diagnostic spine. Serial sections are often needed: only the central slices of a granuloma will contain bits of the egg. In eggs deposited within the previous 4 weeks, the nuclei in the miracidium stain well.

ULTRASONOGRAPHY FOR SCHISTOSOMAL FIBROSIS OF THE LIVER

Workers in the Sudan have found that this is as accurate as wedge resection and histology, and suitable for diagnosis in the field.

Indirect diagnosis

SKIN TEST

There is an immediate sensitivity reaction following within 15 min of the intradermal injection of antigen, usually an extract of adult schistosome worms—a wheal and flare. The test is so insensitive or non-specific (according to the quality of antigen) as to be useless. Many people who have never left Europe give a positive reaction (probably due to contact with avian schistosome cercariae) and many patients with long-established infections are negative.

IMMUNODIAGNOSTIC TESTS

These are numerous. They include the following:

1 CFT.

2 IFAT.

3 Circumoval precipitin (COP) test.

4 The CHR (*Cercarien Hüllen Reaktion*: the development of precipitates around cercaria incubated in a patient's serum).

5 SPAG.

6 ELISA and several others.

Although the better ones correlate well with the results of direct diagnostic methods, they all suffer from the following disadvantages to a greater or lesser extent:

1 They give no indication of the intensity of infection.

2 They do not distinguish between past and present infection.

3 They are not species-specific.

4 Most require high technology.

Immunodiagnostic tests capable of detecting the presence of circulating antigen would be of much greater use to the clinician and epidemiologist, but are still in their infancy.

Circumstantial diagnosis of schistosomiasis

In areas endemic for *S. haematobium*, the presence of haematuria (provided menstruating females are excluded) correlates well with the passage of schistosome eggs in the urine. With a dipstick-type test, provided it can detect both free haemoglobin and discrete red cells (such as the Boehringer product), the number of false positives and false negatives is very low. The false positives will partly be explained by glomerulonephritis, and partly by the passage of dead eggs by patients whose worms are dead.

The radiological changes in the urinary tract may be very suggestive. Almost pathognomonic is the ring-like calcification of the bladder, which may also involve the ureters, prostate and seminal vesicles. The multiple, rounded filling defects produced by pseudopapillomata in the bladder are also very typical.

The presence of colonic polypi in an endemic area incriminates schistosome infection as the most likely cause. So also does the syndrome of portal hypertension with normal liver function tests. Otherwise unaccountable pulmonary hypertension in an endemic area should arouse suspicion of schistosomiasis.

The most useful clue, in cases with disease due to ectopic worms or metastatic eggs, is the presence of eosinophilia. A raised eosinophil count should always alert the physician to the possibility of an invasive helminth infection of some sort, and in this type of case specific immunodiagnostic tests may be useful.

In the initial illness, again the association of eosinophilia with the other symptoms should raise the question of worms. So also should the patient with diarrhoea and eosinophilia, although other worms such as *Strongyloides stercoralis*, *Capillaria philippinensis* and *Trichuris trichiura* can cause the same picture. In the differential diagnosis of these, direct diagnosis by examination of the stools is paramount.

DRUG TREATMENT

There must be an active infection, meaning that living worms must be present. The only justification for treatment without confirmation by direct diagnosis is when CNS involvement is suspected, when the risks of treatment given in error are less than the risks of delay.

Aim of drug treatment

The aim is to kill as many worms as possible without harming the patient. Heavily infected patients may need several courses of treatment to achieve complete parasitological cure. Complete cure is often impracticable in endemic areas, and may be unnecessary provided few living worms are left. Some people think the persistence of a few living worms maintains concomitant immunity.

Test of cure

Worms commonly recover fully from sublethal drug intoxication for up to 6 weeks after the treatment ends, and some may recover for up to 3 months. Tests of cure carried out before 6 weeks after treatment are of little use, as many patients then appear cured whose worms subsequently recover. A 3-month follow-up is ideal.

Drugs

Three drugs have displaced older remedies for schistosomiasis.

METRIPHONATE (TRICHLORFON; BILARCIL)

This is an organophosphorous insecticide active against S. haematobium infection only.

Dosage

Orally 10 mg/kg on three occasions, at intervals of 2–4 weeks.

Toxicity

Negligible (mainly gastrointestinal); results very good. It reduces cholinesterase levels in blood and should be used with care where organophosphorous insecticide exposure is a possibility. Deliberate or accidental overdosage with metriphonate produces a cholinergic crisis requiring urgent treatment with atropine.

OXAMNIQUINE (MANSIL, VANSIL, UK-4271)

A complex tetrahydroquinoline derivative, whose synthesis involves fermentation by an *Aspergillus* mould. It is effective in S. mansoni infections only.

It is less effective in children: doses for sensitive strains are 10 mg/kg twice on the same day. For less sensitive strains, see the package leaflet.

The only common side-effect is dizziness, but the drug may precipitate epilepsy. South American and equatorial African strains are more sensitive than strains from north and south Africa. There is early evidence that even advanced S. mansoni infections with portal hypertension may show regression of symptoms and signs if oxamniquine is combined with supportive measures such as the vigorous treatment of ascites with diuretics.

Preparations

Capsules of 250 and 500 mg. Suspension 50 mg/ml in 30 ml bottles.

PRAZIQUANTEL (BILTRICIDE)
An isoquinoline compound. It is effective against all human schistosomes.

Dosage
40 mg/kg as a single oral dose for S. *haematobium*. 30 mg/kg for two or three doses may be necessary for S. *mansoni* or S. *japonicum* infections.

Toxicity
Giddiness and minor gastrointestinal disturbances. No serious toxicity is reported, but unexplained abdominal pain and short-lived bloody diarrhoea are troublesome in heavy S. *mansoni* infections. It has the most attractive combination of effectiveness, low toxicity and broad spectrum of all the schistosomicides available.

Cost
£1.50 for the average adult treatment makes it too expensive for many developing countries.

DRUG OR VACCINE PROPHYLAXIS OF INFECTION
No practical measure in either direction is in sight. The notion that desensitization to egg antigen would abolish the development of granulomata and thereby enable the infection to be tolerated without ill-effects has been abandoned. However, an irradiated cercarial vaccine has shown promise in preventing bovine schistosomiasis in the Sudan.

Repeated drug treatment—such as periodic administration of metriphonate to entire populations—may be a valuable contribution to disease control. But it is more accurately described as treatment than prophylaxis. Selective chemotherapy of those with heavy S. *mansoni* infections may also be an important way to reduce disease.

EPIDEMIOLOGY

Requirements for transmission
1 Contamination of water with viable eggs from a reservoir host.
2 Presence in the water of susceptible snail intermediate hosts.
3 Suitable environmental conditions for development in the snail.
4 Human exposure to water containing cercariae.

Reservoir hosts

In all three human schistosomes, humans are the main reservoir. Rodents and baboons may be able to maintain *S. mansoni* infection sometimes. Many animals are susceptible to *S. japonicum* infection, including domestic animals such as the horse and dog.

Epidemiological pattern

In most infected communities, infection is commonest and heaviest in children between 10 and 15 years old. Because children have the highest egg output and are most likely to contaminate water, they are usually the most important reservoir of infection.

Exposure may be occupational, such as in workers on irrigated farms and fishermen. Transmission is often focal, and neighbouring villages may have greatly differing endemicities because of this.

Strategy of control

The life cycle can be attacked at various sites, but some sites have proved more vulnerable than others:

1 Contamination of water.
2 Intermediate host.
3 Human contact with infection.

REDUCING CONTAMINATION OF WATER

The main methods used are:

1 Health education.
2 Provision of sanitation.
3 Prevention of access to transmission sites.
4 Reduction of egg excretion by the definitive hosts by treatment (humans) or elimination of the hosts (rodents).

Of all these measures, the one most immediately successful in most circumstances is chemotherapy.

ATTACK ON SNAILS

Permanent results are possible if the habitats can be eliminated. It has been achieved in Japan and many parts of China by drainage and landfill.

Temporary results are obtained with the application of poisonous chemicals (molluscicides) to snail habitats. If used alone, this method is usually disappointing. The number of infected snails may not be reduced in proportion to the total snail reduction.

Disadvantages include cost, the need to reapply chemicals for an indefinite period, and undesirable effects such as killing fish.

It is most effective in highly controlled environments, such as irrigated agricultural estates, and when used in combination with chemotherapy.

REDUCING CONTACT WITH INFECTION

The necessity for contact can be reduced by the provision of a safe water supply.

This will not prevent recreational or occupational contact. Attempts to fence off transmission sites are usually unsuccessful. Health education is important.

Drugs suitable for mass treatment (see Drug treatment, pp. 235–6)

Metriphonate is suitable for *S. haematobium* only. Its advantages are that it is cheap and well tolerated. Its disadvantage is the high defaulter rate with the second and third doses of the traditional three-dose regimen. In some situations periodic single doses may be effective.

Oxamniquine in suitable for *S. mansoni* only. Its advantage is that it is well tolerated. Its disadvantages are its relatively high cost and the need for more than one dose in areas of high resistance.

Praziquantel is suitable for all species. Its advantages are that it is well tolerated and only a single dose is required to achieve significant reduction in egg output. Its disadvantage is the relatively high cost.

QUESTIONS, PROBLEMS AND CASES

1 What is the difference between the incubation period and prepatent period in schistosomiasis?

2 In what group(s) of people is the initial illness most often recognized?

3 A patient in the tropics presents with fever and diarrhoea which continue for 2 weeks. Investigations show the presence of eosinophilia (6000 eosinophils/μl). Describe in general terms the likely cause of this illness.

Make a short list of the helminths which can cause this.

4 A 10-year-old boy in tropical Africa presents with painless haematuria.
   (a) List the two most common causes.
   (b) What question would you ask to help to distinguish between these two causes?

(c) What laboratory examination would be most helpful in deciding the diagnosis?

(d) If schistosome eggs are found, would cytoscopy be helpful?

(e) If you were working in a teaching hospital with an immunology laboratory, do you think any immunological test would help with the diagnosis, if schistosome eggs have been found?

5 In a sophisticated hospital, your investigations of a 15-year-old boy with *S. haematobium* show the presence of a left-sided hydronephrosis and hydroureter. Renal function tests are normal. What would you do next?

6 In a similar West African patient seen in England, the pyelogram repeated 3 months after treatment showed no change. What important test should be carried out before assuming that the uropathy was irreversible?

7 You are asked by the government of an east African country to advise on how to reduce the risk of schistosomiasis transmission in a series of irrigated sugar estates. The project is still in the planning stage. Make a list of your recommendations.

FURTHER READING

Homeida, M.A. *et al.* (1988) Effect of antischistosomal chemotherapy on prevalence of Symmers' periportal fibrosis in Sudanese villages. *Lancet* **2**, 437–40.

Jordan, P., Webbe, G. (1994) *Schistosomiasis: Epidemiology, Treatment and Control.* Heinemann, London.

Prescott, N.M. (1987) The economics of schistosomiasis chemotherapy. *Parasitol Today* **3**, 21–4.

Savioli, L., Mott, K.E. (1989) Urinary schistosomiasis on Pemba Island: low-cost diagnosis for control in a primary health care setting. *Parasitol Today* **5**, 333–7.

WHO (1993) *The Control of Schistosomiasis.* Technical Report Series 830. WHO, Geneva.

Oriental liver flukes

There are three main species which commonly infect human beings. Their morphology, life cycles and pathogenic effects are so similar as to make the distinction between them largely academic.

PARASITES

All the oriental flukes are hermaphrodite creatures of similar appearance, lanceolate in shape, translucent and brownish in colour. In common with other flukes, they possess two suckers. The adults inhabit the bile passages. *Clonorchis sinensis* is about 10–25 mm long by 3–5 mm wide. The other species are about half as big.

Clonorchis sinensis

This is widespread in China and Japan, Korea, Taiwan and Vietnam. The many reservoir hosts include domestic dogs and cats.

Opisthorchis felineus

This is widespread in eastern Europe as a zoonosis of many wild and domestic animals. Human infections are commonest in Poland and Russia.

Opisthorchis viverrini

This is the most common liver fluke in Thailand. In the north, prevalence rates of more than 50% are common. There is a large reservoir of infection in fish-eating animals, including dogs and cats.

LIFE CYCLE IN GENERAL

All three have closely similar life cycles, the differences involving mainly the intermediate hosts. The eggs are small, brown and operculated. They pass out of the body in the faeces of the definitive host, having entered the gut in the bile. On reaching fresh water, they can only develop further if they are eaten by a susceptible species of operculated snail. In the snail's body the miracidium develops into a sporocyst and then a redia, both asexual replicative stages. The mature redia contains many cercariae with unforked tails, which escape from the snail into the surrounding water when the redia bursts.

When they come into contact with a susceptible species of fish (more than 80 species can be infected with *C. sinensis*), the cercariae force their way between the scales, lose their tails and encyst as infective metacercariae. Humans usually become infected from swallowing the metacercariae in raw or slightly pickled fish, although crayfish have been found to be infected in China. The infection cannot develop in sea fish.

The snail–fish cycle takes about 8 weeks, from infection of the snail to the formation of infective metacercariae.

The metacercariae excyst in the duodenum and ascend the bile passages directly. They develop into mature flukes in about 4 weeks.

CLINICAL PICTURE

Of the many millions of people infected by these parasites, only a minority suffer significant illness. As with all other non-replicative helminths, the pathology is related to the number of parasites present. A few, maybe up to 50 or more, cause little trouble. A large load, such as 500 or more, may cause serious disease. A total of 21 000 worms have been recovered from one individual at autopsy.

The pathology — fibrosis around retained eggs, fibrosis and inflammation around the bile ducts, focal ectatic dilatations of the biliary tree and metaplasia of the biliary epithelium — does not give much indication of the usual symptoms. These are the most common:

1 A feeling of something moving about in the liver. This was the commonest symptom complained of by several hundred patients with *O. viverrini* infection seen by the author (DRB) in Thailand.

2 Bouts of fever associated with enlargement and tenderness of the liver, sometimes with jaundice. When associated with eosinophilia this is presumably due to the worms themselves.

3 Similar symptoms associated with polymorphonuclear neutrophil leucocytosis when bacterial cholangitis is presumably superadded.

4 Painless progressive obstructive jaundice with enlargement of the gallbladder, due to flukes impacted in the common bile duct.

Advanced cases may develop biliary cirrhosis, oedema and cholangiocarcinoma. Even in Thailand, hepatoma is a far more common liver tumour.

DIAGNOSIS

This is by finding the characteristic eggs in the stools, using a concentration technique if necessary. But in patients with complete obstructive jaundice, eggs may not be able to pass into the gut, and only appear when the obstruction is relieved. In such cases, percutaneous needling of the liver may reveal eggs in the aspirated bile. The procedure is not recommended because of the dangers of biliary peritonitis and haemorrhage. There is no effective indirect diagnostic method.

TREATMENT

The advent of praziquantel has made all other drugs obsolete. The dose is 20–30 mg/kg twice daily for 3 days. Tolerance is excellent.

EPIDEMIOLOGY AND CONTROL

The disease is self-inflicted in the sense that it can only be acquired by eating raw, infected fish, so humans can easily avoid infection. This is illustrated by the existence of zoonotic infection in many parts of the world where human infection is unknown. However, food habits are very difficult to change, even with vigorous health education. It seems that once you have tasted well-prepared raw fish, the cooked item never tastes as good.

The pungent, fermented fish of northern Thailand, *pla ra*, does not transmit liver fluke infection.

The oriental liver flukes are much smaller than the cosmopolitan *Fasciola hepatica* associated with sheep. A heavy *F. hepatica* infection causes far more constitutional upset, and much more damage to the liver. Blood loss into the bile (haemobilia) is commonly the cause of death in newly infected sheep, and may cause serious anaemia in human fascioliasis.

The metacercariae encyst on various plants, and infected watercress is the usual source of human infection.

The key syndrome is intermittent obstructive jaundice with eosinophilia.

Lung flukes

There are several lung flukes capable of infecting humans, all in the genus *Paragonimus*. The disease is normally a zoonosis, and humans are infected accidentally.

PARAGONIMUS WESTERMANI

This is the most important lung fluke infecting humans. It is widespread in the Far East and South-east Asia. It is also reported from foci in Oceania. The animal reservoir includes wild and domestic felines and canines, pigs, beavers and mongooses.

A very similar parasite, *P. africanus*, causes human disease in west Africa.

The adult worms, which normally live in cavities in the lungs, are stout, brownish-red flukes, about 10 mm long, 5 mm wide and slightly less in thickness. They are hermaphrodites, with oral and ventral suckers.

LIFE CYCLE

The large, operculated eggs are either coughed up or passed in the faeces of the definitive host. When an egg reaches fresh water, it takes several weeks to develop before releasing its miracidium. This then enters the body of certain species of freshwater snail where it grows and replicates as with liver flukes. The cercariae have odd little knob-shaped tails.

The cercariae encyst as metacercariae in various freshwater crustaceans, mainly crayfish and crabs. Metacercariae may be found in the gills, muscles or liver. They are visible as minute, pearly cysts.

Events following infection

Humans acquire the infection by swallowing the metacercariae in raw or slightly pickled crustaceans. 'Drunken crabs' (living crabs immersed in rice wine before consumption) provide perhaps the most picturesque means. The contamination of food preparation vessels with metacercariae is important in some areas.

The metacercariae excyst in the duodenum, and the larval fluke then rapidly penetrates the gut wall and enters the peritoneal cavity. Larvae remain there for some days before migrating through the diaphragm to reach the lungs, where they form nests in the lung parenchyma which communicate with the bronchial tree. Although the flukes are hermaphrodites, they seem to prefer to live in close association with each other. Their eggs are discharged into the bronchial tree and expelled via the ciliary escalator.

Some of the larvae go astray and fail to reach the lungs. These tend to cause the most serious pathology.

Lung involvement

When the flukes reach their normal habitat in the lungs, the symptoms are cough and haemoptysis or cough with rusty sputum. Chest pain is often present, and fever and night sweats may occur in early infections. A diagnosis of pulmonary tuberculosis is often wrongly made.

The 'nests' containing the flukes take the form of one or more cavities, sometimes interconnecting, surrounded by inflammatory infiltration and, later, fibrous tissue. The individual cavities are usually only about 1 cm in diameter, and they can usually only be recognized as cavities by tomography. Retained eggs excite a granulomatous response.

The flukes live a few years. When they die, either naturally or following treatment, they are resorbed and the cavities heal, leaving a fibrous scar. Only in exceptionally heavy and repeated infection is there significant disturbance of lung function.

Ectopic flukes

Flukes developing outside the lungs may cause puzzling and sometimes serious effects:

1 Abdominal pain, signs of peritoneal irritation, and diarrhoea if worms lodge in the gut wall.

2 Painful lymphadenopathy.

3 Migrating subcutaneous swellings with high eosinophilia. Sometimes abscess formation.

4 Blindness if the eye is involved.

5 Inflammation of the epididymis and testis.

6 A variety of serious cerebral symptoms if the flukes reach the brain. This is particularly common in Korea. The condition closely mimics a brain tumour.

DIAGNOSIS

Direct diagnosis

This is by finding the characteristic eggs in the sputum or the stool. In surveys it is usually easier to collect stool specimens than sputum.

The most important aspect of diagnosis is to think of the condition. The clinical picture and radiological findings are often indistinguishable from TB. Paragonimiasis should be especially suspected in patients diagnosed as having pulmonary TB who have no acid-fast bacilli in the sputum smear. It is very easy to miss the presence of *Paragonimus* eggs if the sputum is only examined for *Mycobacterium tuberculosis* using the oil-immersion lens.

The stool should be examined using the direct method and a good concentration technique such as the formol-ether.

The sputum should be examined by direct microscopy, using a $\times 5$–10 objective. Very mucoid specimens are easier to examine if incubated in $\times 2$–3 their volume of 10% potassium hydroxide for an hour first. The deposit produced after centrifuging for 3 min at 1500 rpm is then examined in the usual way. Masses of eggs are often contained in the brown specks sometimes visible in the sputum.

Eggs may also be found in abscess fluid surrounding ectopic worms, and in histological sections.

Indirect diagnosis

This is only of use in cases of ectopic infection where the eggs are inaccessible. The CFT is most widely used, and suffers from the usual disadvantages of all such tests.

TREATMENT

The drug of choice is praziquantel, 25 mg/kg three times a day for 3 days.

EPIDEMIOLOGY AND CONTROL

As with the oriental liver flukes, the disease is acquired by eating raw foods which could be made perfectly wholesome by cooking. Many pickling processes fail to kill the metacercariae. The zoonotic nature of the infection means that environmental sanitation has little relevance to control. The only solution, as with the liver flukes, is to convince people of the need to change their eating habits.

Intestinal flukes

These flukes are mainly parasites of animals, and have a bewildering variety of names. Despite their numbers we shall deal with them briefly, because they are of very little medical importance. The most important is *Fasciolopsis buski*.

FASCIOLOPSIS BUSKI: PARASITE AND LIFE CYCLE

This is a common parasite of humans and pigs in many parts of South-east Asia. The large flattened oval fluke may reach 7.5 cm in length, 2 cm in breadth and 3 mm in thickness. It lives in the upper small intestine, to which it attaches itself by its suckers.

The large operculated eggs cannot be easily distinguished from those of *Fasciola hepatica*. On reaching fresh water, the eggs develop for several weeks before liberating a miracidium which is infective to various freshwater snails. Asexual multiplication in the snail resembles the process in *Clonorchis*, *Opisthorchis* and *Paragonimus* species. The process culminates in the release of cercariae which encyst as metacercariae on the surface of a variety of water plants. Humans are infected by swallowing the metacercariae when peeling edible water plants with the teeth, mainly the water caltrop and the water chestnut.

Human *Fasciolopsis* infection

When swallowed, the metacercaria excysts in the duodenum and remains there. It takes about 3 months to grow into an adult worm capable of laying eggs. Most infections are entirely asymptomatic, apart from the occasional appearance of a worm in the faeces, often following a bout of diarrhoea of unrelated cause.

Very heavy infections may lead to damage to the small bowel sufficient to cause symptoms: diarrhoea, oedema (probably due to protein-losing enteropathy) and even intestinal obstruction. A toxic state is also described. A serious attempt to demonstrate defective small intestinal function in infected children in Thailand was entirely unproductive. Reports of serious effects in the literature are usually ancient.

Diagnosis

Direct smear examination of the stool is sufficient to reveal any significant infection, as the flukes are very prolific. The daily egg production is about 25 000 per worm.

There is no indirect diagnostic method, nor is there any need for one.

Treatment

Praziquantel, as a single dose of 15 mg/kg, is the drug of choice.

Epidemiology and control

The infection is most abundant where pig and human faeces are allowed to enter the sites used for cultivation of edible water plants. Control is a matter mainly of health education, for the metacercariae are killed almost instantly on contact with boiling water.

OTHER INTESTINAL FLUKES

Echinostoma species

Several species can infect humans. The flukes are much smaller than *F. buski*. Reservoir hosts include rodents and dogs.

Most human infections are acquired from eating raw snails. Most cases are seen in South-east Asia.

Heterophyes heterophyes

This is a very small fluke, infecting rodents, canines, felines and even bats and birds. Infection is acquired from raw fish. Cases are reported from China, Japan and Egypt.

Metagonimus yokogawai

This is the smallest intestinal fluke, most common in the Far East, especially Japan. Reservoir hosts include cats, dogs, pigs and pelicans. Infection is acquired from eating raw fish.

All these conditions are diagnosed by finding characteristic eggs in the faeces, all respond to praziquantel, and all could be avoided if people would abandon the dangerous practice of eating raw flesh.

The filarial worms

Wuchereria bancrofti

The infection is confined to humans, and is widely distributed in the warmer parts of the world, including Africa, Asia, south America and Oceania. More than 90% of all infections are found in Asia, where in some cities the disease is an urban epidemic. The worms are slender, white creatures, the males about 4 cm long, the females 10 cm long and rather stouter. Their normal habitat is the lymphatic system. They live for 10 years or more.

LIFE CYCLE

The worms live mainly in the lymphatics of the groins and axillae. Fertilized females liberate motile larval worms, microfilariae (mf), which escape into the blood circulation. The mf survive for a year or more, during which time they undergo no significant change. If they are taken up in the blood meal of a suitable vector (intermediate host), such as one of the many susceptible species of culicine or anopheline mosquito (the culicines include *Aedes* and *Mansonia* spp.), they begin to develop into infective larvae. There is no multiplication in the vector.

Some of the mf in the stomach of the mosquito penetrate the gut wall and reach the flight muscles, where they develop into short 'sausage' forms. After further development, slender filariform larvae are formed, which migrate forwards to the proboscis via the body cavity.

Humans acquire the infection when filariform larvae enter the bite wound while the mosquito is feeding. The time taken for a newly infected mosquito to become infective is at least 10 days, and often much longer.

CLINICAL PICTURE

Early symptoms after infection

The earliest symptoms develop 6 or more months after initial infection, and are due to lymphangitis. The adult worms in the proximal lymphatics draining the limbs and external genitalia cause an intermittent sterile lymphangitis and lymphadenitis. The reason why the attacks are intermittent is not known, nor the exact mechanism whereby worms whose presence is normally well tolerated can suddenly cause acute inflammation. The attacks are apparently not caused by the mf, because many American soldiers briefly exposed to infection in the Pacific in the Second World War developed typical symptoms of filariasis without ever developing microfilaraemia. On the other hand, during surveys one commonly encounters people with large numbers of mf in their blood who have never had any symptoms at all.

Filarial lymphangitis

The commonest features are recurrent bouts of fever, associated with heat, redness, pain and tenderness overlying a lymphatic vessel. The local lymph nodes are usually enlarged and tender. If it can be observed (in a fair-skinned patient), the lymphangitis may be seen to spread distally. This is the opposite to the direction of spread in a patient with lymphangitis secondary to a peripheral septic lesion. Also, a search for a peripheral septic focus is unrevealing. The most frequently affected sites are:

1 Leg (thigh and groin).
2 Arm (upper arm/axilla).
3 Breast (in women).
4 Spermatic cord (funiculitis).

NATURAL HISTORY

The attacks tend to recur at irregular intervals. The interference caused to lymphatic drainage commonly causes some temporary lymphoedema, but in the earlier stages at least, this tends to subside between attacks. Each attack usually lasts 2–3 weeks.

As more attacks occur, there is an increasing amount of residual damage to the lymphatic system which does not resolve between the attacks. The first noticeable effect of this is residual swelling of a limb, or a hydrocoele, which does not disappear when the acute attack subsides. In this early stage the swelling is due to fluid only, and can be relieved by elevating the affected part or by the application of a pressure bandage. This insidiously leads on to the next stage.

Irreversible effects of damage to the lymphatic system

The most important, and the most characteristic, is elephantiasis. This is a true overgrowth of the skin and subcutaneous tissues which results from chronic lymphoedema. The legs and scrotum are most often affected, and massive overgrowth of the tissues leads not only to a grotesque deformity, but to an incapacitating handicap. The scrotum may come to weigh more than the rest of the patient, and the weight of the legs be so great as to prevent walking. The arms may also be affected, and the female breast and vulva.

Dilatation of the lymphatic vessels from lymphatic hypertension leads to visible and palpable lymph varices. A diffuse lymphatic varicosity affecting the scrotum causes a chronic weeping condition in which lymph is constantly exuded from the skin—lymph scrotum. Obstruction to the lymphatics of the small bowel leads to the development of varices which may rupture into the renal pelvis. The leakage of this chylous lymph into the urine causes chyluria, in which the urine resembles pink milk: pink from contained red cells and milky on account of the fat droplets.

Complications of damaged lymphatics

The overgrowth of the skin which attends elephantiasis makes it rugose, with an exaggerated pattern of creases and folds. Bacteria causing secondary infection find a hospitable environment in these closely apposed skin folds with their lymph-rich tissue fluids. They tend to multiply and invade the vulnerable tissues.

Repeated attacks of bacterial cellulitis often contribute to further damage to the lymphatic system, and significantly add to the patient's misery. Streptococci are most often involved.

The load-related pathology paradox

With non-replicating helminths, the severity of the pathology is usually related to the number of worms present. But something odd happens in filariasis:

1 In surveys of infected communities, it is often found that those with obvious filarial disease, such as elephantiasis, have no mf in their blood.

2 In addition, those with very high mf counts often have nothing to show for the infection at all.

So there seems to be a negative correlation between what is generally accepted as filarial disease, and direct parasitological evidence of the infection.

But the relation between infection and the pathology described is well established on epidemiological grounds. One sees elephantiasis only in communities which have parasitological evidence of infection (microfilaraemia), and the higher the median mf count in a community, the greater the proportion of patients with clinical disease. Also, if the abundance of the infection in the community is reduced by mass chemotherapy, the number of patients with elephantiasis and hydrocoele diminishes, to vanishing point in successful programmes.

The paradox is probably explained by the presence of an enhanced immune response to the parasites in those with symptomatic disease. This might lead to the trapping or destruction of liberated mf, or in the later stages of the disease, to the destruction of the adult worms.

DIAGNOSIS

Direct diagnosis

Direct diagnosis depends on finding the mf in the blood. They are sheathed, and when appropriately stained the arrangement of their nuclei allows them to be distinguished from other mf.

The most commonly used method is the Giemsa-stained thick blood film, using a measured drop, usually of 20 μl. In all surveys the mf are counted, so that not only is the prevalence rate measured, but the mean microfilarial density can be calculated.

Various concentration methods are used which allow the detection of microfilaraemia at levels much lower than are shown up by the thick blood film.

Among them are membrane filtration, in which the blood is filtered through a synthetic membrane and the whole membrane stained and examined after mounting in a clearing agent. At least 50 times as much blood as contained in a thick film can be examined in this way. But this requires a venous blood specimen, and is more expensive and time-consuming than a simple blood film.

The number of mf in peripheral blood usually varies greatly throughout a 24-h period. This greatly affects the best time to take blood for examination. Many patients with symptomatic infections have no mf in their blood at any time.

MICROFILARIAL PERIODICITY

Most *W. bancrofti* infections show marked nocturnal periodicity. This means most of the mf appear at night, with a peak between 2200 and 0100 h. During the day the mf are very scanty or cannot be found at all on ordinary blood films. The difference between the maximum and minimum counts is often more than 100-fold. When not circulating in the blood, the mf are sequestered in the lungs. They can be flushed out by giving a single dose of 50 mg of diethylcarbamazine (DEC).

The Pacific strains of *W. bancrofti* show diurnal periodicity, more mf being found in the blood in the day than in the night. Because the difference between maximum and minimum counts is usually 10-fold or less, the periodicity is described as diurnally subperiodic.

Not surprisingly, microfilarial periodicity is synchronized with the biting activity of the appropriate vectors. This nice evolutionary adaptation can be upset by reversing the activity pattern of the host. If a man with nocturnally periodic *W. bancrofti* goes on to night duty, in a few days his mf will be found in his blood in greatest numbers in the daytime, provided he does not sleep on duty.

Indirect diagnosis

This is really only needed where filariasis is suspected, but mf cannot be found in the blood. It can be very useful in patients with elephantiasis of unknown cause. Antibodies may be detected in the blood using the usual array of methods. Most common are the CFT, IFAT and ELISA tests. A skin test using extract of adult worms can also be used: the development of a wheal 15 min after the intradermal injection is positive and indicates the presence of antibody.

All the indirect antibody tests suffer from the usual disadvantages of lack of sensitivity or specificity, the two requirements always being in opposition to each other. A factor in causing false-positive results is that in many tropical countries the infective larvae of animal filarial worms, such as the dog heartworm, *Dirofilaria immitis*, are frequently being injected into people by the mosquito vectors. Even though not usually capable of developing into mature worms, they may give rise to antibody formation. The cross-reactivity between antibodies to the different filarial worms is very great. *D. immitis* is the usual source of antigen used for the CFT for *W. bancrofti*. Cross-reactions with other nematode infection such as *Strongyloides stercoralis* also occur.

TREATMENT

Diethylcarbamazine

There is only one drug widely used at present, diethylcarbamazine (DEC), although some recent drugs show promise. DEC rapidly kills the mf. It also kills the adult worms if given in a prolonged course as follows: full dose 2 mg/kg body weight, three times a day for 3 weeks.

For an adult this usually corresponds to three 50 mg tablets three times a day. It is usual to give 0.5 mg/kg on the first day, 0.5 mg/kg three times on the second day, and then double the dose every day until the full dose is reached.

The drug itself is extremely well tolerated, but it commonly causes reactions in patients with filariasis, due to the liberation of antigens from the mf in the early days of treatment, and to reactions around the dead worms later. The early reactions are fever, malaise, headache, nausea, vomiting and sometimes urticaria. Later there may be lymphangitis, funiculitis, painful lymphadenopathy and abscess formation around the dead worms.

The severity of the initial reaction is reduced if treatment is started with a small dose as described. If reactions are severe, the drug should be stopped for a day or 2 and then restarted at a lower dose. Unless a full course is completed, the adult worms will probably not be killed.

An antihistamine drug given with aspirin will give some relief from the early symptoms. No medicine can prevent reactions to the adult worms. If abscesses form they must be drained.

Ivermectin (see Chapter 32, p. 268) causes prolonged suppression of mf production, but its effects on the disease process are still unknown.

Treatment of the complications

Established elephantiasis may be treated surgically. In the case of the limbs, treatment is very unsatisfactory. Surgical treatment of scrotal elephantiasis is very successful, and it should always be possible to save the testes. The penis, even if completely buried, will be found to be normal when excision of the redundant scrotal tissue reveals it. Hydrocoele is treated surgically in the normal way.

Whilst compression bandaging can reduce the size of the limb it is rarely tolerated in hot environments. Coumarin (5,6-benzo-α-pyrone) used long-term is a symptomatic oral treatment for lymphoedema. DEC treatment has little effect on established elephantiasis, but is usually given in the hope it will prevent further deterioration.

If recurrent streptococcal infections of the skin are a problem, phenoxymethyl penicillin (penicillin V) can be given daily by mouth indefinitely.

EPIDEMIOLOGY

There are two different epidemiological types:
- Urban filariasis. The main vector is *Culex quinquefasciatus*, a night-biting

mosquito which breeds in very polluted water. It is a major problem in some Asian cities.

2 Rural bancroftian filariasis is transmitted by a considerable number of culicine and anopheline mosquitos. In both situations, the disease may cause significant morbidity, but it seldom causes death.

Outline of control measures

1 Attack on mosquitoes. As most *C. quinquefasciatus* are resistant to many insecticides, residual spraying is of little use. Most effective are draining of breeding sites and larviciding. Some antimalarial spraying programmes may have helped reduce transmission in villages.

2 Personal protection by use of mosquito nets (obviously no help in the Pacific, where the vectors bite in the day).

3 Periodic mass treatment with DEC to reduce the infective pool of mf. A suitable regimen is 2 mg/kg three times a day, once a month for a year. The population must be warned of the unpleasant effects, which are less severe after the first two or three doses. This may abolish clinical disease in whole communities, and they may remain free of disease for years, even though transmission has not been entirely stopped. It has been very successful in the Pacific.

QUESTIONS, PROBLEMS AND CASES

1 A 30-year-old African farmer presents with swelling of one leg of about 1-year's duration. List the most likely causes.

2 A 17-year-old youth in a village in South-east Asia presents with a 1-week history of a painful swelling in the right groin. On examination you find a mass of enlarged and tender inguinal glands.
   (a) What questions would you ask him?
   (b) What particular physical signs would you look for?
   (c) In the light of the above findings, list the possible causes.

FURTHER READING

Casley-Smith, J.R., Wang, C.T., Zi-hai, C. (1993) Treatment of filarial lymphoedema and elephantiasis with 5,6-benzo-α-pyrone (coumarin). *BMJ* **307**, 1037–40.

WHO expert committee on filariasis fifth report (1992) *Lymphatic Filariasis; The Disease and its Control*. Technical Report Series 821. WHO, Geneva.

Brugia malayi

Malayan filariasis is much less widespread than bancroftian filariasis. It has a patchy distribution in India, South-east Asia generally, and many of the islands in the Malay archipelago. It is not usually an urban disease, and the infection is both less common and less serious than bancroftian filariasis. The subperiodic strain of the parasite has an animal reservoir in the Kra monkey, and several other animals may also be infected.

PARASITE AND LIFE CYCLE

The adult worms closely resemble *Wuchereria bancrofti* and some skill is needed to tell them apart. The mf are sheathed, but can easily be distinguished from those of *W. Bancrofti* by the characteristic arrangement of their nuclei in stained preparations. The vectors are usually members of the species *Mansonia*, although several *Anopheles* species transmit the infection in towns. In all important particulars the life cycle resembles that of *W. bancrofti*.

CLINICAL PICTURE

The basic pathogenesis is the same as that of *W. bancrofti*, and the clinical features are similar but less severe. The elephantiasis of Malayan filariasis is usually confined to below the knees, and scrotal involvement is not so gross. Chyluria is rare.

DIAGNOSIS

Microfilarial periodicity

The most important diagnostic method is by finding the mf in the blood, as described in Chapter 29, p. 254. The mf associated with infections acquired in open swamp areas are nocturnally periodic, but in swamp forest in the jungle, with an animal reservoir, nocturnally subperiodic. There are no visible differences between them.

Indirect diagnostic methods can be used, but are of little practical importance.

CHEMOTHERAPY

Chemotherapy of B. *malayi* infection resembles that of W. *bancrofti* (Chapter 29, p. 255) and the only difference is in the greater susceptibility of the adult worms to DEC. Doses of 2 mg/kg three times a day for 10–14 days are often sufficient. Reactions to treatment are common.

EPIDEMIOLOGY AND CONTROL

Mosquito control measures are more feasible when the vectors are anophelines. *Mansonia* spp. in their larval stages obtain their oxygen from the stems of emergent water plants. They do not come to the surface of their breeding sites, so cannot be destroyed by applying suffocating oils.

Usually, the only practical control measures in rural areas where the disease is most endemic is by periodic mass chemotherapy, and encouraging the use of mosquito nets. In several parts of South-east Asia, the enthusiastic application of these methods has significantly reduced disease.

Loa loa

The disease is called loiasis, and is confined to forest areas in west and central Africa. The adults are slender, white creatures 3–7 cm long, and infection is probably confined to humans. A closely related parasite is found in monkeys. On account of its limited distribution and relatively mild clinical effects, it is the least important of the pathogenic filarial infections.

LIFE CYCLE

The adult worms inhabit the subcutaneous tissues, in which they move about freely. The fertilized female worms liberate mf into the circulation, where they show diurnal periodicity. mf are taken up in the blood meal of various biting flies in the genus *Chrysops*. They develop in the fly in a similar way to the larvae of bancroftian filariasis in mosquitoes, and finally appear in the proboscis as infective (filariform) stages. They do not replicate in the fly.

It takes about 1 year after inoculation of infective larvae for the worms to become mature. Some patients with obvious symptoms of loiasis never have mf in their blood. Perhaps in light infections the male and female worms have difficulty finding each other.

The mf probably live a year or two, the adults very much longer. DRB has seen a patient with active disease due to living worms 18 years after her last possible exposure to infection.

CLINICAL PICTURE

The wanderings of the worms usually produce no symptoms. But at times painful or itchy swellings suddenly develop in the subcutaneous tissues, to subside a few days later. These are the Calabar swellings, named after a town in eastern Nigeria.

The pathogenesis of Calabar swelling is not known, but perhaps some trivial blow to the body surface injures the worms sufficiently to make them release irritant substances. Calabar swellings are usually 10 cm or more in diameter and poorly demarcated from the surrounding tissues. Their frequent occurrence on the arms supports a relationship to injury.

The most dramatic symptoms of loiasis develop when a worm passes across the eye under the conjunctiva. The patient suddenly complains of 'something in the eye', and this is soon followed by intense irritation, pain and swelling of the periorbital tissues. The patient looking in the mirror is horrified to see the worm, moving with a vigorous thrashing movement and visible between the bulbar conjunctiva and the sclera. The worm usually moves out of the conjunctival area within half an hour or so, but it takes a day or two for all the symptoms to subside.

Worms which die usually calcify in the tissues without doing any harm. They may be seen in disproportionately large numbers in the hands and feet, but this is probably not a reflection of their true distribution. Sometimes the death of a worm gives rise to a chronic, painful, granulomatous lesion.

Loiasis very rarely causes serious complications. It is one of the commonest causes of eosinophilia without symptoms in those who have lived in an endemic area.

DIAGNOSIS

Blood taken during the daytime is examined for the presence of the distinctive sheathed mf, as for bancroftian filariasis (Chapter 29, p. 254). There is almost always a considerable eosinophilia, and serological tests such as the *Dirofilaria immitis* CFT may be positive.

Many patients with a clear history of Calabar swellings and even of worms traversing the eyes have eosinophilia only. Treatment of these patients on clinical grounds is fully justified.

TREATMENT

Surgical removal

The worms can be removed surgically if they can be accurately located, most easily when they are traversing the conjunctiva. After anaesthetizing the conjunctiva with cocaine or amethocaine eyedrops, the bulbar conjunctiva over the worm can be picked up with fine forceps, snipped with fine scissors, and the worm gently extracted intact with forceps.

Although the operation gives enormous satisfaction to the doctor and the patient, there are usually many worms elsewhere in the body, and the removal of one worm is unlikely to cure the patient.

Chemotherapy

The mf and the adult worms are killed with DEC, which is given in the same dose as for bancroftian filariasis (Chapter 29, pp. 255–6). Reactions are fairly common, and mainly take the form of fever, headache and malaise. They usually settle within a few days of the first full dose of drug being given, but after that the dead worms may give rise to tender subcutaneous swellings which take months to subside. Relapse is unusual if the full dose is given for 3 weeks. After treatment the eosinophil count may rise very high (sometimes to more than 20 000/μl) but usually returns to normal in 3 months if treatment has been successful.

Patients with very high mf counts may suffer sudden, severe encephalitic symptoms while being treated with DEC. The main symptom is alteration of consciousness, but focal neurological signs may occur as well. Starting treatment with a very small dose will minimize but not entirely eliminate the risk and deaths have occurred. For this reason it is probably unwise to treat asymptomatic Africans who will continue to live in an endemic area. Steroids are not apparently beneficial. Apheresis to remove the mf from the blood has been used before DEC treatment, with apparent success.

Ivermectin 400 μg/kg body weight gives prolonged suppression of microfilaraemia, but a beneficial effect on symptoms is uncertain.

EPIDEMIOLOGY

The *Chrysops* breeds in association with forest streams, and the flies (females) bite in the daytime, but will not feed in open sun. The infection is a rural disease, as flies seldom bite more than 200 m from their breeding sites. They are attracted to woodsmoke and will bite indoors.

PREVENTION

Control of the vectors and avoidance of bites are impracticable measures in endemic areas.

The most promising control method is regular prophylactic chemotherapy with DEC. A dose of 200 mg three times a day for 3 days a month has been suggested.

FURTHER READING

Martin-Prevel, Y., Cosnefroy, J.Y., Tshipampa, P., Ngari, P., Chodakewitz, J.A., Pinder, M. (1993) Tolerance and efficacy of single high-dose ivermectin for the treatment of loiasis. *Am J Trop Med Hyg* **48**, 186–92.

Onchocerca volvulus

Onchocerciasis is caused by the nematode *Onchocerca volvulus*. Although present in central and south America and the Yemen, 95% of all cases are found in Africa. The disease is found in a wide band across equatorial Africa, extending from the savannah regions in the north to the rain forest in the south. As the disease is a major cause of blindness, the WHO and the countries most affected are cooperating in an attempt to eradicate the infection from huge areas of west Africa.

PARASITE AND LIFE CYCLE

The parasite is found only in humans. Adult worms (males up to 4 cm, females up to 50 cm long) live free in the subcutaneous tissues or in fibrous nodules beneath the skin. They live for many years, during which the females liberate large numbers of minute mf which pass into the skin.

Microfilariae in the skin are taken up by the small biting fly *Simulium*, whilst feeding. *Simulium damnosum* is the vector in west Africa, and *S. naevei* in east Africa. The different behaviour of these species accounts for many differences in the epidemiology of the infection in east and west Africa. The central American flies may be brightly coloured or have a metallic sheen.

Microfilariae develop inside the fly, and after development in the flight muscles migrate forwards to the mouthparts. They escape from the mouthparts during feeding, enter the bite wound, and so infect humans. The time taken from ingestion of the mf to the time when the fly becomes infective is usually some weeks.

Infective larvae develop into adult worms in about 1 year. When adult males and females meet, the female is fertilized and begins to liberate larvae. Sometimes the worms become enclosed in fibrous nodules, possibly because migrating worms in the subcutaneous tissues are temporarily trapped between bony prominences and the hard surface on which the patient is sleeping. This might immobilize the worms long enough for the local tissue defences to imprison the worms in a cellular reaction. This would explain why the nodules are most commonly found over bony prominences such as the greater trochanter, the iliac crest and the lateral chest wall.

If the adult worms are killed, the mf live in the skin for between 1 and 2 years. The adult worms can live and reproduce for up to 20 years.

CLINICAL PICTURE

Effects on the skin

The pathogenesis of *O. volvulus* is almost entirely due to the mf, and the adult worms normally cause no symptoms. Nodules are no more than a minor cosmetic blemish, and migrating worms not confined in nodules cause no trouble.

Microfilariae cause pathology when they die, and form a small focus of inflammation. This reaction varies greatly, but in the skin it initially causes itching. Itching is commonly followed by a papular rash, aggravated by consequent scratching. Secondary infection of excoriations may be severe. Often the itchy skin looks normal, especially when the infection is light.

More severe skin disease is produced by numerous mf, and the skin changes in Africa are usually most severe in the lower half of the body. This is almost certainly because the flies bite in the lower half of the body and more adult worms are found in the lower half of the body than in the upper. The number of mf in the skin is related to the distance from the worms which produce them, so the mf density in the lower half of the body is usually higher than in the upper half.

Prolonged heavy infection leads to irreversible structural changes in the skin. The skin loses pigment and depigmentation often takes the form of small spots of pigment surviving in depigmented areas, most commonly in the shins. There is also damage to the elastic tissue of the skin, leading to an appearance typical of old age. Wrinkles form, and skin sometimes becomes so redundant as to hang down. When this change is extreme and associated with hypertrophy of the inguinal lymph glands, the 'hanging groin' is produced. More usually the premature ageing of the skin (presbydermia) is what first strikes the eye. It is most obvious on the lower limbs, although the thorax and upper arms may be involved in advanced cases.

Enlargement of the inguinal glands often occurs, sometimes associated with minor elephantiasis of the legs or scrotum. Onchocerciasis is not an important cause of elephantiasis and never causes gross elephantiasis.

Effect on the eyes

If the pathology of onchocerciasis were confined to the skin the disease would not have attracted much attention. But mf can enter the eye, sometimes leading to blindness. As blindness is most commonly found near rivers in which the *Simulium* flies breed the name river blindness arose in Africa.

The pathology of the eye is divided into two main parts — that in the anterior segment and that in the posterior segment. In the anterior segment, the earliest lesions are due to mf in the cornea. Small collections of inflammatory cells in the cornea lead to a typical punctate keratitis which has been described as 'snowflake keratitis' because of the appearance of a little group of white opacities together, which resemble the appearance when a snowflake strikes a window. Punctate keratitis usually resolves and does not normally lead to corneal scarring. The more severe pathology is of vascularization and organization of the cornea, starting at the lower and lateral aspects of the corneoscleral junction and gradually spreading upwards and inwards in such a way eventually as to obscure the pupil. This sclerosing keratitis, often associated with iritis caused by the death of the mf in the iris muscle, is the major cause of blindness in onchocerciasis.

A less common cause of blindness is posterior lesions.

Retinal damage occurs, with or without optic atrophy. The view that posterior segment changes are due to genetic factors has been abandoned, and their relationship to onchocerciasis is now generally accepted. Further eye complications are mainly consequent on those already described. Most important are secondary cataracts resulting from synechiae, and glaucoma.

SYMPTOMS

The commonest symptom of onchocerciasis is itching. There are no visible lesions in fairly mild infections apart from the presence of excoriations. In severe cases, the itching may be so intolerable that the patient cannot sleep at night, and it has even been known to lead to suicide.

The earliest symptoms of eye involvement are redness and irritation, sometimes associated with lacrimation. At this stage there are often no visible lesions. When the eyes are severely affected the patient commonly complains of photophobia and more florid lacrimation. As the pathology progresses the vision becomes gradually reduced. Where the lesions are mainly of the anterior segment, there is first blurring of vision, then restriction of vision to the upper half of the visual field due to the ascent of the sclerosing corneal lesions. Eventually this leads to virtually complete blindness, and when corneal opacification is complete, the patient may not be able to perceive light.

When blindness is due to optic atrophy, the patient may not notice impairment of vision until constriction of the visual fields has caused gross tunnel vision.

Special features of onchocerciasis in central America

In central America the flies often bite on the head, and nodules on the scalp are common. Diffuse reddish swelling of the facial tissues, erisipelas de la costa, may result from intense mf infiltration, especially in children. Depigmentation may follow in adults.

Natural history of untreated onchocerciasis

Mild infections may continue for many years without causing any symptoms other than minor itching, and the mildest infections may be entirely asymptomatic. In general, only when there is repeated reinfection over a period of two or more decades do serious eye complications develop. There are many parts of tropical and subtropical Africa where the disease is endemic but the rate of blindness is very low. On the other hand, there are parts of Africa where the transmission rate is so high and the mf densities so large that complications in the eye can be seen before the patient reaches the age of 20 years.

There is some evidence that African forest strains of the parasite are less pathogenic to the eye than are savannah strains.

DIAGNOSIS

Direct diagnosis

Specific diagnosis is made by finding mf of *O. volvulus*. These are most commonly found in the skin, and this requires that a small piece of skin be removed. The commonest sites for skin snips to be taken are from the lateral aspect of the calf, the thighs and over the hip or iliac crest. In central America, snips can be taken from the outer canthus or shoulder. In research programmes as many as six snips from different sites may be taken to estimate both mf density and the distribution of mf vertically in the body. The higher the mf count in skin taken from near the eye (usually from the outer canthus), the more likely it is that serious eye disease will be present.

TECHNIQUE FOR TAKING SKIN SNIPS

The skin over the selected site is cleaned with alcohol and the alcohol allowed to dry. The tip of a fine-pointed needle is then inserted into the skin and raised so the skin forms a sharp-pointed tent. A sharp sterile blade (razor or scalpel) is taken in the other hand and a piece of skin 1–2 mm square is removed beneath the needle point. No local anaesthetic is needed and the pain caused by this procedure, provided it is carried out rapidly, is trivial. The skin taken should leave a clearly visible white area beneath the epidermis. Into this white area, if the thickness of the snip was adequate, blood will begin to well from damaged capillaries. The skin snip is then transferred into about 0.2 ml of physiological saline solution. This is most conveniently done by using plastic microtitre trays. The snips should be left in the saline solution for 24 h in order to allow mf to emerge from the snip, and strips of adhesive cellulose placed over the wells to prevent evaporation. It is convenient to do several snips one after the other so the advantages of this 'mass production' technique can be exploited to the full.

After 24 h at ambient temperature in the tropics, the saline solution surrounding the snip is aspirated with a Pasteur pipette, placed on a microscope slide, and examined without a cover slip with a ×10 objective. If the snip is incubated for 24 h, approximately five times as many mf will emerge as from a snip incubated for only half an hour. There is no need to tease out the snip, as this does not increase the yield of mf.

If the half-hour incubation is used, all the mf emerging from the skin snip will be alive. If 24-h incubation is used, then some mf will be dead. It is possible, with a little practice, to differentiate the mf of *O. volvulus* from other mf that may be present in the skin snip, particularly the smaller, hook-tailed mf of *Mansonella streptocerca* (not pathogenic) and the sheathed mf of *Loa loa*, or the smaller, unsheathed mf of *M. perstans*, both of which originate from contamination of the skin snip by blood. Until the microscopist has had practice in recognizing *O. volvulus* mf (they have a typically enlarged cephalic end), it is wise to stain some of the mf with Giemsa or similar stain so their detailed morphology can be compared with the illustrations in a parasitology textbook.

The number of mf in the skin snip should be recorded as this gives an indication of the intensity of infection. When expressed in terms of mf/mg of skin—a technique used in research surveys and when chemotherapy is being evaluated—the highest counts may exceed 1000 mf/mg. The normal size of a good skin snip is somewhere between 1 and 4 mg. If mf are found in skin taken from the outer canthus, then there is great likelihood that mf will be found in the eye. The method described uses only simple equipment which should be found in any hospital. A much more convenient method uses the corneoscleral punch forceps of

Walser. These are usually only available for research projects as they are expensive and require resharpening after a few hundred snips have been taken.

Microfilariae may also be seen moving actively in the anterior chamber of the eye by slit-lamp microscopy. The patient should first put his or her head between the knees for 5 min to bring mf into view.

Other diagnostic methods

Serology is not yet widely available for the diagnosis of onchocerciasis, and there is no useful skin test. Eosinophilia is usual, but common to many invasive worm infections. Onchocerciasis is a common cause of asymptomatic eosinophilia in a patient who has lived in an endemic area.

The Mazzotti test can be used when mf have not been found. If mf are killed by the drug DEC, a brisk reaction occurs in the host even at levels of mf density undetectable by conventional methods. The technique is to give the patient 50 mg DEC by mouth. If the test is positive, the patient will notice either the aggravation of existing pruritus or the appearance of pruritus for the first time, will often develop fever, and may develop more severe reactions such as oedema affecting the limbs and face, a papular rash, postural hypotension, vomiting and adenitis. This test must only be carried out on patients with negative skin snips and no eye lesion, or there is a risk of serious reaction and loss of visual acuity.

TREATMENT

Ivermectin

The treatment of onchocerciasis has been revolutionized by the arrival of ivermectin. This drug, a chemical modification of avermectin, one of a series of naturally occurring macrocyclic lactones produced by *Streptomyces avermitilis*, destroys the mf without producing a severe Mazzotti reaction. But, unlike DEC, it has a prolonged suppressive effect on mf production by acting on the reproductive system of the adult female worms. It acts very much like the contraceptive pill in making the worms temporarily infertile. The duration of the effect, after a single oral dose of 150–200 µg/kg body weight, extends up to a year. However, the drug has no lethal effect on the adult worms, so the dosage has to be repeated annually until the adult worms have died of old age. It is not yet known how long one must continue annual medication, but it may be longer than 10 years. In heavy infections ivermectin may have to be given more often, such as 6 monthly, to keep mf levels low.

By an act of extraordinary generosity, the manufacturer, Merck, Sharp and Dohme have agreed to supply the drug free to any government wishing to use it as part of an onchocerciasis control programme. The drug is available as scored tablets of 6 mg. The usual adult dose is two tablets. The only significant toxic effect is the Mazzotti reaction.

The arrival of ivermectin had made the use of DEC and suramin obsolete.

Nodulectomy

The most obvious way of getting rid of the adult worms where nodules are present is to remove them. This method has been applied successfully in parts of central America where most of the nodules are around the head, and where it seems that most of the worms live in nodules. In most parts of Africa the nodules are usually around the lower part of the body and it seems that only a proportion of worms live in nodules. Removing the nodules is a simple procedure, but it must be realized that only a proportion of the worms will be removed. All the same, if a nodule is present near the head, its likelihood of infecting the eyes with microfilariae is so high that it is justifiable to remove it.

EPIDEMIOLOGY AND CONTROL

The distribution of *Onchocerca* infection in Africa is determined by the presence of the *Simulium* fly which is the vector of the infection. In east Africa, the infection has been eradicated from some places, such as the mountainous area around Kericho in Kenya, because the local vector (*S. naevei*) does not travel far. Many years after the streams draining the Kericho plateau were treated with insecticide there has been no reinvasion of the area with *Simulium*, despite the fact that *Simulium* is known to be breeding within 80 km.

The situation in west and central Africa is quite different, for *S. damnosum* has an enormously long flight range, and flies have been found 300 km from the nearest breeding site. They obviously do not fly so far under their own wing power, and it seems they are swept upwards in the air currents, transported along in high-altitude winds and then deposited where they are found. Forms of life behaving in this way have been described by biologists as aerial plankton. It is this particular characteristic of *S. damnosum* which makes the eradication of the infection from the most severely affected parts of central and western Africa a very great problem. It is sad that the most severe human infections are found beside breeding sites such as the large river systems in western Africa where the potential for agricultural development by irrigation is greatest.

This has impelled the WHO, with the aid of the west African countries involved, to mount an eradication programme using the technique of spraying larvicides, first temephos (Abate, an organophosphorous compound) and later *Bacillus thuringiensis* into the major and minor river systems of the states involved, mainly Ivory Coast, Burkina Faso, Ghana, Togo, Dahomey, Niger and Nigeria. The WHO realized that because the human population has not been rid of mf, the spraying programme would have to last for 20 years to allow the mf in the skin of infected patients to disappear following the death of adult worms from old age. But the situation has now been revolutionized by the availability of ivermectin for mass treatment, with the prospect of eliminating the reservoir of infection.

FURTHER READING

Abiose, A., Jones, B.R., Cousens, S.N. *et al.* (1993) Reduction in incidence of optic nerve disease with annual ivermectin to control oncherciasis. *Lancet* **341**, 130–4.

WHO expert committee on oncherciasis, third report (1987) Technical Report Series 752. WHO, Geneva.

Dracunculus medinensis: *guinea worm infection*

Dracunculus medinensis is the largest of the filarial worms afflicting humans, the adult female being sometimes more than I m long. It is an ancient scourge described in the Bible and by Hippocrates. It is widely but patchily distributed in the Middle East, Arabia, tropical Africa, Asia and parts of south America.

PARASITE AND LIFE CYCLE

Humans and a variety of animals are infected naturally by swallowing small copepod crustaceans of the genus *Cyclops* in contaminated fresh water. The infective larvae in the body of the *Cyclops* are liberated by digestive juices, penetrate the gut wall, develop deep in the body, then take up residence in the subcutaneous tissues. Male worms are seldom found, and the female worms cause the pathology.

The female worm takes about a year to mature, at which time her body is almost entirely occupied by the uterus, filled with rhabditiform larvae. The head end of the worm approaches the skin, where a blister forms. When the blister bursts the uterus prolapses through the skin, and discharges the larvae on immersion in water. When swallowed by *Cyclops*, the larvae penetrate the gut and develop into infective forms in about 4–6 weeks.

PATHOGENESIS

The pathogenesis is caused by allergic and inflammatory response to the adult female in the tissues. About 90% of worms emerge from the legs, but they can emerge from anywhere, especially the arms, genitalia and breasts. Worms entering joint cavities may cause arthritis.

Secondary infection may cause cellulitis, septicaemia or tetanus.

SYMPTOMS

The symptoms, as the female worms become mature, are:
I Localized pain.

2 Swelling.

3 Itching.

4 Fever.

5 Urticaria.

   More severe, but less common symptoms, are:

I Diarrhoea.

2 Vomiting.

3 Wheezy dyspnoea.

4 Postural hypotension.

   When the local blister bursts, symptoms usually rapidly improve.

Characteristic local lesion

A papule develops over the end of the worm, soon followed by a blister. The blister bursts spontaneously, following much pain. The pearly loop of prolapsed uterus can be seen in the centre of the small ulcer which remains.

DIAGNOSIS

The appearances described are diagnostic. But the larvae can be demonstrated if the ulcer is immersed in water, or if a drop of water is placed on the ulcer. This will cause reflex contractions of the uterus and the expulsion of larvae. The motile larvae are 500–700 μm long, with a characteristically pointed tail.

Indirect diagnostic methods are unreliable.

TREATMENT

There has been little improvement since the traditional method was described by Hippocrates.

I The uterus of the worm is emptied of larvae by repeated exposure of the ulcer to clean water. This is done by immersing the foot in a bowl of water for half an hour, or applying wet compresses to less accessible ulcers for a similar period. This is normally done daily for a few days, until no more larvae (visible to the naked eye as a milky fluid) emerge from the worm.

2 The worm will now have emerged from the ulcer sufficiently to be pulled out a little way by forceps. It should be attached to a sterile orange-stick or small piece of wood either by a thread or by jamming the worm in a cleft in the stick.

3 The worm should be gently drawn out of the ulcer by rolling it round the stick until resistance is felt. If excessive force is used the worm will

break, and the leakage of its body contents into the tissues will cause a severe inflammatory reaction. There is a greater danger of this happening if extraction is attempted before the uterus has been emptied. After each attempt at extraction, normally once a day, the ulcer is carefully protected against secondary infection by applying a sterile dressing.

It usually takes about 2 weeks to remove the worm intact in this way, sometimes longer.

Surgical removal of the guinea worm through small transverse incisions can greatly shorten the period of disability.

Drug treatment of guinea-worm infection

Metronidazole 400 mg twice daily for 5 days has been reported to shorten the time taken to extract the worms. But it appears to do this by an anti-inflammatory action, and there is no evidence of direct effect on the worms. In particular, drugs seem to have no effect in preventing the emergence of worms immature at the time of treatment. In poor developing countries the relatively high cost makes drug use unjustified.

Tetanus prophylaxis should be given.

What happens to worms that are not removed?

The indigenous population of endemic areas is often well aware of the Hippocratic method, and native healers are often proficient at removing worms, but secondary infection often occurs because of ignorance of aseptic principles.

Worms left *in situ* will usually die and calcify, and the local wound usually heals without permanent sequelae. Calcified worms are often seen in various parts of the body on routine radiological examination. But the removal of worms, using a careful technique as described, will materially shorten the period of invalidity and reduce the likelihood of complications.

EPIDEMIOLOGY AND CONTROL

The disease is associated with the lack of a clean, perennial water supply. It is often very seasonal in transmission, for during the dry season people may have to rely on very contaminated surface water such as that produced by excavating a hole in a river bed.

Any water supply that the people can stand in while collecting water is a danger. Most notorious are the step wells of India.

New infections can be prevented instantly by teaching people to sieve their drinking water. Two layers of ordinary shirting material are adequate to remove *Cyclops*, on account of its large size. Advice to boil drinking water is usually impracticable owing to the scarcity and high price of fuel in tropical countries.

Another short-term solution is to kill the *Cyclops* in the water supply by adding Abate (temephos). The long-term solution to the problem is a safe water supply.

The eradication of guinea-worm infection is entirely feasible and great progress has been made in its eradication from the Indian subcontinent whilst extensive eradication campaigns are in progress in many African countries. The WHO has named guinea-worm infection as a candidate for global eradication.

Tropical eosinophilia syndrome

DEFINITION AND CAUSE

Tropical eosinophilia syndrome is the name given to a condition characterized by pulmonary symptoms and high eosinophilia, believed to be caused by cryptic filarial infection. The disease has a similar distribution to bancroftian and Malayan filariasis, but is far more often recognized in Asia than anywhere else. It has been produced experimentally in humans infected with the filarial parasite of animals, *Brugia pahangi*. It is probably more often a hyperallergic response to *Wuchereria bancrofti* or *B. malayi*.

Children and young adults are affected most frequently, especially in the Indian subcontinent.

CLINICAL PICTURE

The onset is gradual, sometimes with lassitude and low fever. The main symptoms are cough, often nocturnal, and wheezy dyspnoea. The signs indicate bronchospasm and lung-function tests confirm an obstructive picture. The chest X-ray may be entirely normal, or show a variety of abnormalities, including:

1 Generalized loss of translucency.
2 Slight hilar adenopathy with perihilar loss of definition and blurring of the vessel margins.
3 A coarse interstitial pattern.
4 Interstitial nodules.
5 Small patches of consolidation (rarely).

There is a very high eosinophilia, with an eosinophil count often more than 20 000/μl.

DIAGNOSIS

Microfilariae are virtually never found in the blood. The most useful test is the *Dirofilaria immitis* CFT, which is usually strongly positive. The diagnosis is confirmed by the response to treatment.

Differential diagnosis

The differential diagnosis is from other worms that migrate through the lungs and may cause respiratory symptoms and signs. The most important one is *Ascaris*, but gross patchy infiltrates are usually present in all these conditions, and they cause a short self-limiting illness rather than a chronic disability.

TREATMENT

The response to DEC is rapid and complete: X-ray changes usually return to normal within a week, and lung function is restored a few weeks later. Subjective improvement occurs in a few days. The dose is the same as for *W. bancrofti*.

CONTROL

Control of the condition depends on the control of filariasis.

QUESTION

I You have a young Asian city-dweller with what sounds like asthma of recent onset. His chest X-ray is normal, his eosinophil count 12000/μl and his blood film negative for mf. You cannot get the *Dirofilaria* CFT done. What would you do next?

Anaemia in the tropics

CHAPTER 35

Anaemia in the tropics

Anaemia complicates many tropical diseases. The cause of anaemia is usually, if sufficiently precisely defined, the actual diagnosis. But to define the cause, when it is not immediately obvious, really involves two logical stages. The first stage is to categorize the anaemia into one of several recognizable types. The second stage is to proceed to identify the cause.

CATEGORIES

The whole process can be reduced to a simple algorithm, by asking questions in a logical sequence, and only proceeding to the next question when the preceding one is answered 'yes' or 'no'. The answers are provided by relevant laboratory investigations. To deal with categorization first, these are the crucial questions to be answered:

1 Is the patient anaemic?
2 Is the anaemia hypochromic?
3 If hypochromic, are there numerous target cells and is there gross anisocytosis and poikilocytosis?
4 If red cells are normochromic, are they normocytic or macrocytic?
5 If normocytic, is there a raised reticulocyte index?
6 If macrocytic, is the anaemia megaloblastic?
7 If megaloblastic, is the anaemia due to vitamin B_{12} deficiency?

To answer these questions, some or all of the basic investigations described later will have to be carried out. It is illogical to ask for all the investigations from the outset, for the category of anaemia may be revealed before the whole series of questions has been completed. For example, a hypochromic microcytic anaemia will not be due to B_{12} or folate deficiency, and it would be only logical to ask for serum folate and B_{12} levels if there were some special additional reason for doing so, such as the presence of a malabsorption state, or the observation of large numbers of hypersegmented neutrophil polymorphs in the blood film.

Physical signs

Apart from routine examination, there are several observations that should be deliberately sought and carefully recorded:

1 Is there pallor of the skin in a person normally pigmented? Very dark-skinned people who develop severe iron-deficiency anaemia sometimes complain that their skin has become pale or 'blond'. This occurs in iron deficiency of any cause, but is most pronounced when there is an accompanying hypoalbuminaemia. This combination is a characteristic feature of HWA.

2 Is there koilonychia of the hands? Koilonychia of the toes is not, by itself, a helpful physical sign. In people who habitually go barefoot, more than half of them may have big toes with concave nails in the absence of anaemia. Koilonychia of the hands is a good indication of iron deficiency, but its absence does not exclude it.

3 Are the palms of the hands—in a patient who is normally pigmented such as a dark-skinned African or Indian—black? Striking pigmentation of the palms may occur in Asians and Africans with megaloblastic anaemia, whether due to folate or B_{12} deficiency.

4 Is the spleen enlarged? As splenomegaly occurs in a range of tropical infections, such as malaria, kala-azar and schistosomiasis, the observation of splenomegaly may be helpful.

5 Is the liver enlarged? If so, is it tender? The liver is often slightly enlarged in malaria; often considerably enlarged in *Schistosoma mansoni* and *japonicum* infections, and may be enormous in visceral leishmaniasis. If the liver is very tender, in the absence of a raised jugular venous pressure (JVP), then the possibility of a liver abscess should be considered. The anaemia of amoebic liver abscess is usually only moderate.

6 Is there a fever? Fever may reflect the cause of the anaemia, as in malaria, or it may merely be an accompaniment of the anaemia, as often occurs in severe megaloblastic anaemia of any cause.

7 Is the JVP raised?

8 Is the blood pressure low?

9 Is the patient breathless at rest?

(These last three questions give an indication of how dangerous the anaemia is to the immediate prospects of survival. Other grave prognostic signs are hypothermia and mental confusion. These signify reduced oxygen supply to the tissues to such an extent that life is threatened.)

10 Are there retinal haemorrhages?

Basic investigations

These include:

1 Accurate Hb estimation.

2 The haematocrit reading (packed cell volume, PCV).

3 Examination of a well-made thin blood film stained at pH 6.8 with a good Romanowsky stain such as Giemsa or May–Grunwald.

4 Only if a red cell counting machine is available—the red blood cell count (RBC). Manual counts are usually so unreliable that they are not worth doing.

5 The reticulocyte index (RI). This is the reticulocyte count expressed as a percentage of red cells which are reticulocytes, but corrected for the degree of anaemia. By using this correction factor, the discriminating value of the count in indicating bone marrow activity is increased to a worthwhile extent. The correction is simple:

$$RI = \text{reticulocyte count (\%)} \times \frac{RBC \ (\text{millions}/\mu l)}{4.8}$$

if the RBC has been done. If it has not been done, the PCV is used to correct the figure, so:

$$RI = \text{reticulocyte count} \times \frac{PCV}{45}$$

This is not quite so accurate, because the correction factor will be affected by red cell size. In practice, this error is of little importance.

6 A thick blood film stained for malarial parasites where relevant. In inexperienced hands, a thin film (stained by Giemsa or Leishman's stain at pH 7.2) is more reliable. But very low parasitaemias are then easily missed, and the film should be examined for at least 20 min.

FURTHER INVESTIGATIONS THAT MAY BE NEEDED

1 Bone marrow smear for the appearance of the blood cell precursors, parasites and stainable iron.

2 Serum iron, iron-binding capacity and saturation.

3 Serum folate.

4 Serum B_{12}.

5 Serum ferritin.

(But as these last four estimations can only be done in relatively sophisticated laboratories, we will describe later how simpler methods can be used to obtain similar information in developing countries.)

6 The sickling test, or the sickle haemoglobin solubility test.

7 The measurement of fetal haemoglobin, most simply carried out using the alkali denaturation test of Singer.

8 Haemoglobin electrophoresis.

9 Testing for occult blood in the stool.

10 Microscopy of the stool, and if hookworm eggs are present, an egg count.

This rather long list of tests shows what the laboratory may be asked to do. In developing countries, although many of these tests will not be available, the application of a little ingenuity will ensure that the standard of diagnostic accuracy is only slightly affected.

Haematological parameters

This is the numerical information that can be derived from basic estimations: Hb level, haematocrit (PCV), and, when available, RBC. The parameters are:

1 The MCHC:

$$\text{MCHC} = \frac{\text{Hb (in g/dl)}}{\text{PCV (\%)}} \times 100$$

It signifies the percentage of red cell cytoplasm that is Hb. The normal range is 32–36%. Between 30 and 32% it suggests hypochromia; below 30%, definite hypochromia. When the MCHC is well below 30%, hypochromia shows on the thin film. Minor degrees of hypochromia are detectable by a reduction in the MCHC before the red cells look pale.

2 The MCV:

$$\text{MCV} = \frac{\text{PCV (\%)} \times 10}{\text{RBC (millions/}\mu\text{l)}}$$

The result is in femtolitres (fl), 1 fl being 10^{-15} l. Values below 80 fl signify microcytosis; above 100 fl, macrocytosis.

3 The mean corpuscular Hb:

$$\text{MCH} = \frac{\text{Hb (g/dl)} \times 10}{\text{RBC (millions/}\mu\text{l)}}$$

The result is in picograms (pg), 1 pg being 10^{-12} g. The MCH does not really add to the information given by the MCHC and MCV. The normal range is 27–31 pg.

DIAGNOSTIC PROCESS IN ANAEMIA: A SIMPLE SCHEME USING SIMPLE METHODS

Sorting out the categories

We now have the basic information to approach the logical categorization of anaemias. We need to go through the basic seven questions we started with, and see how the system works. The abbreviated scheme is printed below, so it can be seen as a whole before considering each question in some detail.

1 IS THE PATIENT ANAEMIC?

Ask for Hb estimation, PCV, MCHC and blood film. (The simple copper sulphate screening test may be used first if laboratory facilities are limited.)

2 IS THE ANAEMIA HYPOCHROMIC OR NORMOCHROMIC?

Judge from the MCHC (normal over 30%) and the appearance of the film.

3 IF HYPOCHROMIC, ARE THERE NUMEROUS TARGET CELLS?

No: the cause is iron deficiency. Other features include low RI, low serum iron, absence of stainable iron in the marrow.

Yes: the patient has thalassaemia. This is confirmed by an exaggerated degree of anisocytosis and poikilocytosis.

4 IF NORMOCHROMIC, IS IT NORMOCYTIC OR MACROCYTIC?

Judge from the blood film and the MCV if an automatic red cell counter is available (normal MCV = 78–94 μm^3). (Manual RBCs are so inaccurate as to be a waste of time.)

5 IF NORMOCYTIC, IS THERE A SIGNIFICANT RETICULOCYTOSIS?

Judge by a RI of over 2%.

No: there is marrow aplasia or depression.

Yes: there is blood destruction (haemolysis) or blood loss (haemorrhage). (Always do the sickling test and think of G6PD deficiency in unexplained haemolytic anaemias—and do not forget the possibility of malaria.)

6 IF MACROCYTIC, IS THE ANAEMIA MEGALOBLASTIC?

Judge from the presence of hypersegmented polymorphs (more than 30% with five or more lobes) in the blood film. Megaloblasts (and giant metamyelocytes, which are easier to recognize) are present in the bone marrow—but marrow examination is best left to the expert.

No: the cause may be liver disease, haemolysis, myxoedema, aplastic anaemia, etc.

Yes: the cause is folate or vitamin B_{12} deficiency.

7 IF MEGALOBLASTIC, IS IT DUE TO FOLATE DEFICIENCY?

Judge from the reticulocytosis after 200 μg folinic acid intramuscularly if serum folate and B_{12} levels cannot be determined. Only folate-deficiency anaemias will respond to such small doses (B_{12} deficiency anaemia will respond to big doses of folic acid). Failure to respond to 50 μg of B_{12} intramuscularly also indicates folate deficiency.

The questions again: in more detail

1 IS THE PATIENT ANAEMIC?

The detection of anaemia by clinical observation is not easy unless the Hb is less than 10 g/dl, and even then it can be missed in dark-skinned people, in whom the buccal mucous membrane and nails may be pigmented. A screening test, to be successful, must be more accurate than clinical guesswork. Unfortunately, the simplest screening test, using the famous Tahlquist paper, is so inaccurate as to be virtually useless. Tested by African medical students and compared with the cyanmethaemoglobin method used with an accurate colorimeter, the author (DRB) found the Tahlquist papers gave a result within approximately -50% to $+100\%$ of the true value. So if the Tahlquist paper reads 7 g/dl, all one can say is that the true value lies between 3.5 and 14 g/dl. This is not an improvement on guessing.

But the copper sulphate specific gravity screening method is worthwhile, especially when serum protein levels are not grossly abnormal.

Short of photoelectric colorimetry, the Lovibond haemoglobinometer is probably the most reliable, simple method currently available for Hb estimation. But we think it is best used for screening only, because it is a stepwise method, and does not give continuous readings. Greater accuracy is needed if the following stages of the discrimination process are to be reliable.

2 AND 3 IS THE ANAEMIA HYPOCHROMIC OR NORMOCHROMIC, AND IF HYPOCHROMIC ARE THERE NUMEROUS TARGET CELLS?

Hypochromia means poor haemoglobinization of the red cell stroma. It is indicated by a low MCHC and, in grosser cases, by the blood film. The normal red cell has about one-third of its total diameter occupied by the central 'pale area'. Where the pale area is more than one-half of the cell diameter, the cell is said to be hypochromic. The ability to make these observations reliably depends on experience, especially as the appearance of the red cells varies greatly in different parts of the film. The MCH is also low in hypochromic anaemia, and is a much more sensitive indicator than the MCHC when an automatic red cell counting machine is available.

Hypochromic anaemias

Most hypochromic anaemias are due to iron deficiency, and iron-deficiency anaemia is the most important type of anaemia in the world. The red cells typically show only moderate anisocytosis and poikilocytosis, the RI is less than 2%, the marrow lacks stainable iron and the serum iron level is low (below 70 μg/100 ml or 13 μmol/l). However, there are other causes of hypochromic anaemia.

Thalassaemia

This is most commonly due to congenital inability to synthesize Hb β chains. It produces a blood picture very like iron-deficiency anaemia, but there are some important differences. These are:

(a) There are often many target cells.

(b) There is usually a much greater degree of anisocytosis and poikilocytosis than seen in iron deficiency.

(c) The RI is slightly raised rather than depressed.

(d) The marrow contains abundant stainable iron.

(e) The serum iron is normal or raised.

(f) With improvements in diagnostic methods, α-thalassaemia is increasingly often incriminated as the cause of mild hypochromic anaemia.

As patients with thalassaemia often need repeated blood transfusion, they readily develop haemosiderosis which, involving the heart, is often the cause of death. There may be considerable enlargement of the spleen in thalassaemia, and this may contribute, via secondary hypersplenism, to accelerated red cell destruction and hence aggravation of the anaemia. But the spleen may also enlarge because it becomes a source of extramedullary erythropoiesis. Obviously, splenectomy will only benefit patients with hypersplenism.

Skeletal changes associated with marrow proliferation—especially thickening of the diploic spaces in the skull, bossing and the typical 'hair-on-end' appearance of the X-ray—are often seen in thalassaemia. They also occur in the haemoglobinopathies in which abnormal haemoglobin chains are assembled. Identical changes are sometimes seen in chronic iron-deficiency anaemia in childhood. In all these afflicted children there may be exaggerated growth of the long bones.

Hypochromic anaemia and infection

Chronic severe infection may produce a hypochromic anaemia like that of iron deficiency. But this anaemia does not respond to iron and the serum iron level, while depressed, is accompanied by a depression of the total iron-binding capacity (TIBC). Also, stainable iron may be present in the bone marrow.

Pyridoxine deficiency

This also causes a hypochromic blood picture, but it is rare and significant only in infants fed on milk deficient in pyridoxine. Reputable manufacturers of milk for infant feeding have long since put this right.

Mean corpuscular volume and iron deficiency

This is definitely useful in diagnosing iron deficiency, for hypochromia is often preceded by microcytosis, so the detection of microcytosis may lead to an earlier diagnosis. But this parameter can only be derived if an accurate RBC is available.

The anaemia is not hypochromic

In practice, this means that the anaemia is normochromic, for there is no type of anaemia in which the concentration of Hb in the red cells is more than normal—which would, by analogy, be called hyperchromic. Red cells may lose water in hyperosmolar dehydration and so become hyperchromic, but this is temporary and unrelated to anaemia.

4 IS THE NORMOCHROMIC ANAEMIA NORMOCYTIC OR MACROCYTIC?

This question is answered by the red cell size. It is most accurately determined by the MCV, but macrocytes can be recognized in the blood film after a little practice. This stage of division of the normochromic anaemias is vital, as the two groups comprise completely separate aetiologies.

Normochromic normocytic anaemias

There are only two broad causes of this type of anaemia:
(a) acute blood loss or blood destruction (as in haemorrhage and haemolysis);
(b) marrow depression, as in renal failure, various infections, radiation and chemical poisoning.

5 IF NORMOCYTIC, IS THERE A SIGNIFICANT RETICULOCYTOSIS?

The crucial distinguishing feature between the two groups is the level of red cell production. This will be increased in (a) and reduced in (b, above). It is measured most simply by the RI, although in advanced laboratories elaborate ferrokinetic methods give more reliable results. But to illustrate a typical case, a patient with a normochromic normocytic anaemia due to a gastrointestinal haemorrhage may have an RI of 8% or more. A patient with the same blood picture due to chloramphenicol aplasia may have an RI of 0.1%.

In the case of acute haemolysis due to malaria, there is an important *caveat*: in the very early stages of the anaemia, there may be marrow depression or dyserythropoeisis produced by the infection itself. In such cases, reticulocytosis may be absent, and in consequence the haemolytic origin of the anaemia may be obscured. But this will be only temporary, being followed in a few days by a reticulocytosis.

6 IF MACROCYTIC, IS THE ANAEMIA MEGALOBLASTIC?

Macrocytic anaemias are virtually always normochromic. So the patient has large, normally haemoglobinized red cells. But they are inadequate in number, accounting for the anaemia and the disproportionately low RBC.

Most important are megaloblastic anaemias due to deficiency of vitamin B_{12} or folate. In the thin blood film, the presence of many hypersegmented polymorphs indicates this. In fact, even in a hypochromic anaemia, the presence of 30% or more polymorphs with five or more nuclear lobes is a sure sign that there is a concomitant megaloblastic state. In some such cases there is dimorphism—apparently two separate populations of circulating red cells: some small and hypochromic, others large and normochromic.

Strictly speaking, megaloblastic anaemia can only be diagnosed by finding megaloblasts, either in the bone marrow or in the blood. Although megaloblasts can be identified by any careful observer with a little practice, most tyros find giant metamyelocytes, also typical of megaloblastic anaemia, much easier to recognize.

If there is a normochromic anaemia without megaloblastosis, and the serum B_{12} and folate levels are normal, cirrhosis may be the cause. But there is need for caution here: an increase in MCV is one of the earliest signs of alcoholism, and is apparently a direct effect of alcohol on the red cell or its precursors.

7 IF MEGALOBLASTIC, IS THE ANAEMIA DUE TO VITAMIN B_{12} OR FOLATE DEFICIENCY?

The straightforward approach to this question is to measure the serum B_{12} level (normal 160–925 µg/l). If the level is normal, then folate deficiency can be assumed to be the cause by exclusion. And in sophisticated laboratories B_{12} and folate levels are often estimated together. But there is a rather complex relation between B_{12} and folate levels which does not necessarily throw light on which is the important deficit. One way of avoiding these complications is therapeutic trial, but this must be done using physiological doses. Not only can large doses of folic acid produce reticulocytosis and at least partial haematological remission in cases of B_{12} deficiency, but they may also precipitate rapidly progressive subacute combined degeneration of the cord, with crippling and permanent neurological sequelae.

The recommended therapeutic trial procedure is to give 50 µg hydroxocobalamin intramuscularly, and monitor the RI for 7 days. A reticulocytosis confirms that B_{12} deficiency is the cause of the anaemia. Conversely, the appearance of a significant (5% or more) reticulocytosis after giving 200 µg of folinic acid intramuscularly proves that folate deficiency is the trouble.

QUESTIONS, PROBLEMS AND CASES

In most instances I shall assume only the laboratory resources likely to be available in a district hospital in a tropical country.

1 A 35-year-old African farmer presents with a complaint of weakness and breathlessness of 6 weeks' duration. He looks pale, the Hb is 6 g/dl, MCHC 24% and the blood film shows hypochromic red cells but is otherwise normal. Apart from pallor, there are no abnormal physical findings.

(a) What specific questions would you ask him? (The picture is hypochromic anaemia: you want to identify the cause.)
(b) What does the history suggest so far?
(c) Does the history suggest HWA?
(d) What would be the most useful investigations to do next?

2 An Asian woman aged about 40 years presents with a very long history of weakness and tiredness. She looks pale, her Hb is 5 g/dl, MCHC 36%, and the blood film is negative for malaria parasites. The red cells look normochromic and numerous hypersegmented polymorphs are present.

(a) What do the blood findings suggest?
(b) What questions might help you decide on the most likely cause?

3 A 22-year-old west African woman presents in the eighth month of her fourth pregnancy with symptoms of anaemia. Her Hb is 7 g/dl, MCHC and red cell morphology normal. The blood film shows fairly numerous rings of *Plasmodium falciparum*, and she has a slight fever. The RI is 8%, and the stool contains scanty hookworm eggs.

(a) What is the most likely cause of her anaemia?

(b) How would you treat her?

(c) What medication would you give her for the remainder of the pregnancy and the puerperium?

4 A 4-year-old boy from the Persian Gulf area presents with a profound hypochromic anaemia and splenomegaly. The blood film shows hypochromic red cells, gross anisocytosis and poikilocytosis, and numerous target cells.

(a) What is the most likely cause of his anaemia?

(b) What is the danger of treating such a patient with iron?

(c) What is the prognosis in such a case?

5 A British engineer has returned home from a 2-year tour of duty in Asia with a 3-month history of diarrhoea with pale, bulky, offensive stools. He has lost 8 kg in weight and is obviously anaemic. Routine haematological investigations show an Hb of 8.4 g/dl, MCHC of 34% and MCV of 110 fl.

(a) What is the most likely cause of his anaemia?

(b) Are there any other investigations you should do?

(c) How would you treat him?

6 An African boy of 2 years is brought to your clinic screaming vigorously, with tender fusiform swelling of all the fingers of both hands. He looks pale and the spleen is 5 cm below the costal margin. His forehead looks unduly prominent.

(a) What is the most likely diagnosis?

(b) What are the usual routine haematological findings?

(c) What routine medication should such children receive to keep them in as good health as possible?

(d) In children with sickle-cell disease the Hb is often maintained at around 7–8 g/dl without the need for blood transfusion. If the Hb level suddenly falls to 4 g/dl, what are the major causes?

7 A west African soldier, at routine medical examination, is found to be anaemic: Hb 9.6 g/dl. He denies any relevant symptoms at present or at any time in the past. Routine investigations show a normochromic, normocytic anaemia, with numerous target cells in the blood film. The stool contains scanty hookworm eggs, but is negative for occult blood. The sickling test is positive and the RI is moderately raised at 5%. What is the likeliest cause of his anaemia?

8 A 15-year-old east African schoolboy is referred with a history of fever, weight loss and poor school performance. He is seen to be thin, pale and febrile. There are no localizing physical signs, but the results of investigations are as follows: Hb 9.8 g/dl; MCHC 33%; blood film normal;

WBC and differential normal; no malaria parasites seen; ESR 90 mm/h; chest X-ray normal; blood urea normal; urine examination (on three occasions) — albumin: trace deposit; numerous pus cells; urine culture (by dip-slide) negative.

SECTION 8

Leprosy

CHAPTER 36

Leprosy

Leprosy is a disease of humans caused by *Mycobacterium leprae*, an acid-fast bacillus which only grows well at temperatures lower than the body core. The major pathology is confined to the cooler parts of the body: the skin; the upper respiratory mucosa; superficial nerves; lymph glands; the testes; the anterior chamber of the eye. The organism cannot be grown *in vitro*.

DISTRIBUTION

The disease is found all over the world where poverty and overcrowding encourage its spread. It is not especially associated with a warm climate. It has not decreased in the past three decades, partly because even where prevalence rates have fallen, the world population increase has more than offset this. The WHO prevalence estimates for active disease are as follows: 1965 < 11 million; 1970 > 11 million; 1990 11–15 million. Only 3.7 million of these are registered. The World Health Assembly at its 44th meeting in 1991 set a target for the elimination of leprosy as a public health problem by the year 2000 — that is, a reduction in cases to below 1 per 10 000 people.

TRANSMISSION

This is mainly from person to person. Prolonged intimate contact is not necessary, but increases the risk. Most cases probably result from exposure of a susceptible subject to a relatively large challenge, such as by inhaling the countless bacilli expelled when a lepromatous patient sneezes. Ulcers in multibacillary cases may also shed bacilli. But even contacts of patients with tuberculoid disease, in whom few or no bacilli can be found, have a risk of infection several times higher than the general population. This could be due either to shared exposure, or to infectivity of tuberculoid cases which cannot be demonstrated by any other way.

In highly endemic areas everyone is exposed to infection, yet few develop disease. This closely resembles the situation in TB.

293

INITIAL INFECTION

Because most initial infections produce few symptoms and are followed by spontaneous recovery, only a few are recognized clinically. Many initial infections are entirely asymptomatic. Some people develop transient lesions which, even when recognized, resolve spontaneously. The visible lesion is called indeterminate leprosy, because it is impossible to predict its future evolution. When the progress of indeterminate lesions is followed without intervention, most will be seen to resolve completely, and only the minority will develop clinical disease. The outcome depends on the host's immune response: those with the lowest degree of resistance develop lepromatous leprosy, while those with the highest, but not complete, resistance develop tuberculoid leprosy.

LEPROMATOUS LEPROSY

The process

Bacilli are numerous. They are found in:
1 Skin.
2 Nasal mucosa.
3 Nerves (especially in Schwann cells).
4 Lymph glands.
5 Liver.
6 Muscles.
7 Eyes.
8 Testes.
9 Bone marrow.

The host responds by infiltrating the areas where bacilli are present with macrophages and a few lymphocytes. Bacilli can multiply inside the macrophages. A diffuse infiltrate develops, called a leproma. Antibodies are produced, often in large amounts, but they are ineffective in killing the bacilli. No effective CMI develops. The lepromin test, a measure of CMI, is always negative.

The skin

Skin lesions are diffuse, ill-defined and symmetrical. In dark skins, they are often hypopigmented; in light skins, erythematous. The morphology varies and includes:
1 Diffuse infiltration with generalized thickening of the skin.
2 Loss of eyebrows (madarosis).

3 Macules.

4 Nodules.

5 Papules.

The affected areas of skin may look greasy. The individual lesions blend gradually into the surrounding skin. The lesions are not usually anaesthetic.

The nose

The nasal mucosa is infiltrated with bacilli and cells. Mucus is discharged containing countless bacilli. Ulceration of the mucosa may occur. There may be destruction of the septum and adjacent bone. The vessels of the nasal mucosa seems to act as a filter for circulating bacilli.

The nerves

Nerve twigs in the affected areas of skin are infiltrated, but usually not enough to destroy them. Superficial nerve trunks are commonly involved, which may lead to severe nerve damage. The nerves most affected are:

1 Ulnar.

2 Median.

3 Lateral popliteal.

4 Sural.

5 Facial.

The nerve trunks are typically thickened, partly due to cellular infiltration, partly to oedema. Nerve involvement in lepromatous leprosy is typically symmetrical. Nerve damage affects both sensory and motor function in mixed nerves, but sensation often goes first.

The eyes

Leprosy can damage the eyes in three ways: lepromatous iritis can occur in multibacillary disease, and the cornea may be damaged by a combination of loss of sensation and facial nerve damage.

Other features

There may be swelling of infected lymph glands, which may break down and discharge even in the absence of secondary infection.

Infiltration of the testes may lead to testicular atrophy and gynaecomastia. The diversion of antibody-producing resources to making useless antibodies directed against dead bacilli may, by antigenic competition, reduce the ability to produce antibodies to other infections, much as in visceral leishmaniasis.

TUBERCULOID LEPROSY

The process

There are few bacilli, confined mainly to the skin and nerves. The host responds vigorously by surrounding the bacilli with macrophages transformed into epithelioid cells, and giant cells may develop. The nests of epithelioid cell reaction are demarcated into discrete areas by surrounding lymphocytes. The lesion is localized by the CMI process and, if it spreads, it does so slowly with a tendency of the parts first affected to heal, with varying degrees of residual atrophy or fibrosis.

The skin

The following features are typical:
1 Single lesion, or two or three at most.
2 Hypopigmented in dark skins, red in light skins.
3 Sharply demarcated from surrounding normal skin.
4 Anaesthetic.
5 Raised edge, scaly surface, lack of sweating.
6 Tendency to central healing.
7 Tendency to spontaneous cure.
8 Destruction of hair follicles.
9 Loss of sweat and sebaceous gland function.

The nerves

The nerves in the lesion are usually infiltrated with granulomatous tissue sufficiently to lead to loss of sensation (first to temperature, then to light touch, finally to pinprick) with or without detectable thickening. A named nerve near to the lesion may also be infiltrated and thickened, with loss of function. Usually only one nerve trunk is involved by tuberculoid leprosy. In all cases, nerve involvement is asymmetrical. Occasionally, there may be nerve involvement without a detectable skin lesion.

THE CLINICAL SPECTRUM OF LEPROSY (FIG. 36.1)

This is largely governed by the host's immune response. We owe the rational classification of leprosy, uniting the clinical and histological

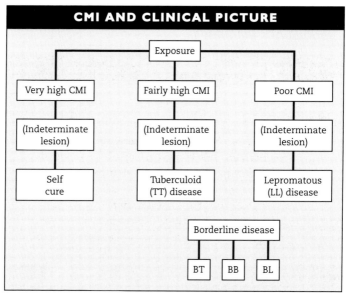

CMI AND CLINICAL PICTURE

Exposure

Very high CMI — Fairly high CMI — Poor CMI

(Indeterminate lesion) — (Indeterminate lesion) — (Indeterminate lesion)

Self cure — Tuberculoid (TT) disease — Lepromatous (LL) disease

Borderline disease

BT   BB   BL

Fig. 36.1   Cell-mediated immunity (CMI) and the clinical picture.

findings by an understanding of immunology, to the work of Ridley and Jopling. Lepromatous and tuberculoid leprosy are merely the opposite poles of a clinical spectrum, which includes great numbers of patients who do not fit properly into either category. Those in this group have borderline leprosy. Those who more closely fit the tuberculoid (TT) category are called borderline tuberculoid (BT), and those who more closely fit the lepromatous (LL) category are called borderline lepromatous (BL). Where the patient's clinical and immunological status seems to be about halfway between the polar types, the disease is classified as borderline borderline (BB) (Fig. 36.2). All these three borderline categories are unstable, in that the patient's immune status may change at any time, and he or she may move towards one or other of the poles. This is an important distinction from the stability of patients with polar disease, who usually maintain their immune status unchanged for an indefinite period.

EXAMINING THE LEPROSY PATIENT

1 Examine the entire body with good illumination for any skin changes. Note texture, colour, sweat gland function, hair distribution and any other abnormality, such as the presence of bone resorption and ulcers.

2 Examine all the accessible superficial nerves for thickening. This

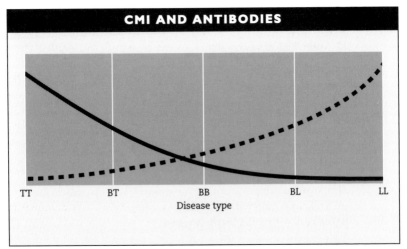

Fig. 36.2 Cell-mediated immunity (CMI) and antibodies. —— CMI status as judged by lymphocyte transformation in the presence of *M. leprae*; - - - - antibody levels.

includes searching for nerves in or near to identifiable skin lesions, a systematic examination of all nerve trunks, and a search for thickened nerves elsewhere. These include all those mentioned, and also:

(a) the back of the neck;

(b) the dorsal surfaces of the hands and feet;

(c) the supraorbital area.

3 Examine for sensory loss and motor weakness. Pay particular attention to the hands (loss of sensation, and weakness and wasting of intrinsic muscles), feet (loss of sensation and footdrop) and face (weakness of orbicularis oculi and corneal sensation).

DIAGNOSIS

Skin-slit smear

This is the most important way of detecting multibacillary disease. In pure lepromatous (LL) cases bacilli can often be found anywhere in the skin. Preferred sites are:

1 Edge of a lesion.

2 Earlobe.

3 Dorsal surface of the first phalanx of a finger.

HOW TO DO A SMEAR

1 Clean the skin with alcohol.

2 Pinch it into a fold, and squeeze it between the thumb and finger to expel blood.

3 Incise the fold of skin with a sterile scalpel blade to a depth of about 1 mm and a length of 5 mm.

4 Turn the blade through 90° and scrape the lips of the wound with the sharp edge of the blade so as to remove some tissue pulp on the point, making sure that the skin remains squeezed so that there is no contamination by blood.

5 Smear the pulp on a clean, new microscope slide, and spread it evenly into an area of 0.5–1 cm^2.

6 Dry in air.

7 Fix with a flame.

8 Stain with modified Ziehl–Neelsen, using only 1% acid–alcohol to avoid decolorizing the bacilli altogether.

READING THE SMEAR

Examine with the oil immersion lens. Bacilli may be seen free or in macrophages as globi. The number of bacilli is recorded on an agreed scale called the bacterial index:

Over 1000 bacilli in an average field 6+.
100–1000 bacilli in an average field 5+.
10–100 bacilli in an average field 4+.
1–10 bacilli in an average field 3+.
1–10 bacilli in 10 fields 2+.
1–10 bacilli in 100 fields 1+.

It is usual to examine 100 fields before describing the smear as negative, i.e. the bacterial index = 0.

The morphological index

This means the proportion of bacilli that look normal and stain as solid-looking rods. When beaded or fragmented, the bacilli are probably not viable. When on effective treatment, the morphological index begins to fall long before the bacterial index falls. It seems to take the body a long time to tidy up the dead bacilli.

Nasal smear

This is done by scraping the inferior turbinates with a long, fine spud or by making the patient blow the nose, and making a smear from the expelled mucus. It gives a good idea of infectivity.

HISTOLOGY

The smears are usually positive in LL, BL and BB cases and, if so, histology is largely academic. On the other hand, in many cases of tuberculoid disease, the combination of the typical raised, anaesthetic, sharply demarcated lesion when associated with thickening of a nerve is clinically diagnostic. Histology is helpful in doubtful cases, and tissue should be taken from the active edge of a lesion. Most information is given by a stain such as Trichrome-fite-faraco (TRIFF), in which not only is tissue morphology shown well, but the bacilli are also stained.

Lepromin test

This is a skin test using leprosy bacilli. A standardized suspension of killed bacilli (0.1 ml) is injected intradermally. The test is read at 72 h (like the Mantoux test) and is called the Fernandez reaction. It indicates delayed hypersensitivity to the antigens in *M. leprae*. A nodule 3 mm or more in diameter is positive. The test is read again at 3–4 weeks (Mitsuda reaction); a positive test now shows that the patient is capable of developing CMI to *M. leprae* or already possessed some CMI when the test was done. Neither test is specific, for other mycobacteria share antigens with *M. leprae*. The test can never be used to diagnose leprosy.

USE OF THE LEPROMIN TEST

Its use is mainly in classification and for prognostic purposes. If a patient with borderline leprosy becomes lepromin-negative, when previously positive, the CMI status is deteriorating. The opposite applies too. Patients with LL leprosy are always lepromin-negative. This remains so throughout life, and is unchanged by treatment.

It is possible to diagnose and treat leprosy patients succssfully without ever using the lepromin test. The growth of *M. leprae* in the nine-banded armadillo has enabled researchers to produce large amounts of purified fractions of the bacilli. None of them has yet been shown to be any better than crude lepromin.

TREATMENT

Principles of chemotherapy

These are to destroy the bacilli and render infectious patients non-infective as soon as possible. A rapid reduction in the population of bacilli in multibacillary cases minimizes the chance of drug resistance emerging, and the principles are exactly the same as those governing the treatment of tuberculosis. It is a major tragedy that leprologists were so slow to accept the truth of this.

In multibacillary cases at least two drugs should be given together, to minimize the danger of drug resistance.

Contrary to ancient teaching:

1 Chemotherapy should never be started with small doses and gradually increased. A full dose should always be given from the start.

2 Effective chemotherapy should never be stopped during reactions of any kind.

Taking into account effectiveness, cost, toxicity, compliance, and the need for treatment of limited duration, the WHO (Technical Report Series 675, 1982) has recommended two basic regimens:

1 MULTIBACILLARY LEPROSY (MBL)

Multibacillary leprosy is defined as all LL, BL and BB patients, and also those BT patients who have one or more smear sites with a bacterial index (Ridley's scale) of 2+ or greater (10 or more acid-fast bacilli seen in 100 oil-immersion fields).

Regimen for normal-sized adults

1 Rifampicin 600 mg supervised monthly.

2 Clofazimine 300 mg supervised monthly plus 50 mg unsupervised daily.

3 Dapsone 100 mg unsupervised daily.

The monthly dose of rifampicin and of clofazimine should be swallowed in front of the doctor or a reliable paramedical worker.

Dosage will depend on body weight. The dose of dapsone is 1–2 mg/kg body weight per day. The dose of rifampicin is 450 mg monthly in patients weighing 35 kg or less, 300 mg monthly if 20 kg or less and 150 mg monthly if 12 kg or less. If the 50 mg clofazimine capsules are unavailable, 100 mg is given on alternate days.

If clofazimine is unacceptable because of its effect on skin colour, the alternative drug is either ethionamide or prothionamide (these two drugs act in the same way, give cross-resistance and are interchangeable). The dosage in adults is 250–375 mg daily, unsupervised as a single dose. Gastrointestinal side-effects are minimized if the dose is taken after the main meal of the day. Hepatitis is a problem in some parts of the world and the combination with rifampicin may lead to fatal and irreversible liver damage.

The triple-drug regimen should be given until the patient becomes smear-negative. Previously untreated LL patients may require 5–11 years, untreated BL perhaps 3–6 years, and untreated BB 2–3 years of treatment. Treated, inactive, smear-negative LL and BL patients (who may be incubating secondary dapsone-resistant infection) should receive 2 years of the regimen.

Relapsed, smear-positive patients, who may or may not have dapsone resistance, should be treated until they become smear-negative (minimum 2 years). Smears should be taken serially from the site of the relapse lesions. The regimen will successfully treat new patients, whether or not they suffer from primary dapsone resistance; relapsed patients, whether or not they have developed secondary dapsone resistance; and old patients apparently successfully treated with dapsone monotherapy, but who may be incubating secondary dapsone resistance. The only variable is the length of treatment. Should a patient relapse after stopping therapy, then no new drug resistance will have been acquired.

2 PAUCIBACILLARY LEPROSY (PBL)

PBL is defined as all indeterminate, TT and BT patients who, when untreated, have no smear site with a bacterial index greater than 1+ on Ridley's scale (<10 acid-fast bacilli seen in 100 oil-immersion fields).

Regimen for normal-sized adults

1 Rifampicin 600 mg supervised once monthly for 6 doses.
2 Dapsone 100 mg unsupervised daily for 6 months.

Should a patient be receiving steroids for a reversal reaction at the end of the 6 months of treatment, the dapsone is continued until the steroids are stopped.

Dangers of dapsone alone in multibacillary disease: resistance

Although most patients treated with dapsone alone will respond well, in some, secondary (acquired) dapsone resistance will develop. This occurs in about 3% of cases per year, and after 10 years of treatment with dapsone alone almost half may be resistant. When acquired resistance in LL cases does develop, lesions of BL type tend to appear. Resistance can be confirmed by growing the bacilli in the footpads of immunosuppressed mice, and showing that normal doses of dapsone are ineffective in curing the infection. Resistance to the other drugs may also develop if prolonged monotherapy is used.

In poor countries, dapsone may have to be used alone on grounds of economy. If so, every effort should be made to reduce the initial bacterial load by giving at least one dose of rifampicin.

'Persisters'

These are dormant organisms which persist in the skin, lymph glands, muscles and elsewhere, even after prolonged treatment with highly bactericidal drugs and drug combinations. They are important in the LL, BL and BB types of disease. There is no way they can be destroyed by drugs, and they are believed to be the main cause of relapse when chemotherapy is stopped.

Reactions in leprosy

These are a source of great confusion. We have put them in the section on treatment because they usually—but not always—develop while the patient is receiving treatment. With one exception (the down-grading reaction) they have one characteristic in common: inflammation. Some reactions are very mild, some so severe as to threaten the patient with death or crippling disability.

ERYTHEMA NODOSUM LEPROSUM (ENL)

This occurs in lepromatous and near-lepromatous disease, at some time or other, in more than half of all patients. It is an antigen–antibody complex disease, and usually develops when most of the bacilli are dead. When immune complex forms, it activates complement and causes acute inflammation at the site of its deposition. In all but the mildest cases there are generalized effects, such as fever, malaise and polymorphonuclear leucocytosis with a raised ESR. The main focal effects are:

1 Lumps: development of red, hot, tender swellings in the skin. They tend to appear in crops at intervals of a few days.

2 Arthritis: varying from mild pains in the joints to changes resembling pyogenic arthritis.

3 Iritis.

4 Neuritis: pain and swelling in a nerve or nerves, with corresponding loss of function.

5 Nephritis: albuminuria.

If the ENL reaction is prolonged and severe, amyloidosis may develop. ENL is sometimes called 'type 2 reaction'.

Treatment

MILD CASES

Bed rest and aspirin in full doses.

MODERATE CASES

Introduce clofazimine, up to 300 mg/day, for its anti-inflammatory effects. Do not continue this dose for longer than 3 months, because of the danger of abdominal complications. Use aspirin as well.

SEVERE CASES

Bed rest and a corticosteroid such as prednisolone 40–60 mg/day in divided doses, rapidly reducing to the minimum dose that will control symptoms. If symptoms are not controlled on a dose that does not cause signs of hypercorticism, change to thalidomide, 100 mg four times daily. This must never be used in women of child-bearing age. The effect of thalidomide in this condition is dramatic and sometimes life-saving.

ENL usually subsides by itself in 6 months or less. It should not be necessary to continue thalidomide for longer than this, but it is sometimes far harder to wean patients off steroid drugs.

Antileprosy treatment should be continued throughout the ENL reaction.

Reversal reactions

There is slight disagreement about the use of this word. We think it is logical to use it to describe any reaction in which the immune status of the host changes, one way or the other. Some use it to describe a reaction due to an increase in CMI only. What we think is the most logical concept is shown below.

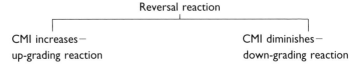

Reversal reaction

CMI increases –   CMI diminishes –
up-grading reaction   down-grading reaction

They are both mainly complications of the unstable borderline types of disease.

UPGRADING 'REVERSAL' (TYPE 1) REACTIONS

CMI increases, so lesions which were previously not inflamed suddenly increase in size, become red and inflamed, and are usually sharply divided from the surrounding tissues. A nerve affected by this process suddenly gives rise to pain, and there may be a very rapid loss of function. Nerve abscesses may form and pressure in the nerve sheath may aggravate the damage done by inflammation. The lepromin test, if previously negative, may become positive.

Treatment

There is no time to hesitate, because nerve damage may be rapid and severe. Prednisolone 40–60 mg/day should be given, and improvement is usually very rapid, within a few days. If there is sudden swelling of a nerve associated with pain and loss of function which does not respond within 24 h to high-dose steroid treatment, the nerve should be decompressed. This is done by:

1 Freeing the nerve from any constricting tissue (e.g. dividing the carpal ligament).

2 Incising the nerve sheath longitudinally.

More extensive surgical intervention is unhelpful and may be harmful.

DOWN-GRADING REACTIONS

These only occur in untreated (or incompletely treated) patients. Disease which was previously BT typically begins to deteriorate as judged by:

1 A tendency to develop new lesions.

2 The appearance of bacilli in smears.

3 The conversion from Mitsuda-positive to negative.

All these consequences follow on a loss of CMI, and the unchecked proliferation of the bacilli. One can summarize these effects by saying that untreated borderline leprosy is immunologically unstable, and often becomes suddenly worse in the absence of treatment.

Treatment

This simply involves the institution of effective antileprosy chemotherapy. Acute inflammation plays no part in this type of reaction (in fact, it is lack of reaction that is the main feature of the condition), so immunosuppressive treatment is not indicated. It would, in fact, be harmful.

Interaction between steroids and rifampicin

Rifampicin is a powerful inducer of liver enzymes. In patients receiving rifampicin daily, the liver inactivates steroids with much greater speed than normally. So the normal dose of steroids may be ineffective, and about twice the usual dose may be needed. When rifampicin is stopped, the dose of steroid can usually be substantially reduced.

Treating other complications of leprosy

ANAESTHESIA OF EXTREMITIES

The most important measure is health education, to teach the patient to

avoid injury, mainly burns of the hands and friction damage to the feet.
The provision of suitable shoes is a difficult problem.

PENETRATING ULCERS OF THE FEET

Prevention is the aim. Ulcers develop over pressure points such as the
metatarsal heads and calcaneum. Excision of the metatarsal heads helps
to spread the weight-bearing area. When ulcers will not heal, or repeatedly break down with conservative treatment, they will often heal if
weight-bearing can be prevented by fitting a below-knee plaster cast with
a weight-bearing hoop.

DAMAGE TO THE EYE

If this results only from inability to close the eye, a temporalis transfer
operation should be tried.

If there is also loss of corneal sensation, some form of tarsorrhaphy
is better.

FOOTDROP

Although a tendon transplant may succeed, a toe-raising spring is much
simpler.

DAMAGE TO INTRINSIC HAND MUSCLES

Specialized operations are available to restore the lumbrical function lost
from ulnar nerve palsy, but they are only successful where the joints are
mobile and good aftercare is available.

Transplanting the flexor tendon of the third or fourth finger to the
thumb may restore opposition where median nerve palsy has led to its
loss.

EPIDEMIOLOGY

Facts

The modes of transmission are still not understood, and the importance
of genetic factors in resistance remain unproved. But we do have a few
facts available:

1 Human beings are the only natural reservoir of infection (rarely,
armadillos).

2 The organism replicates relatively slowly.

3 Known contact with any case of leprosy increases the risk of developing the disease.

4 Contact with a lepromatous case produces a higher risk.

5 The disease may not become manifest until 10 years after contact.

Figures

Endemicity is described as follows: 10 cases per 1000 population—hyperendemic; 1 case per 1000 population—leprosy is an important public health problem.

In some highly susceptible communities previously unexposed to infection, as many as 30% of the population may eventually become infected.

The WHO estimates that transmission tends to stop when the prevalence rate is less than 5 per 10 000.

School-age children provide a convenient indication of endemicity: the overall population prevalence is about four times as high as the number of cases in the children.

Control strategy

Leprosy prevalence tends to diminish in endemicity as social and economic conditions improve, but this information is not much use to countries with a leprosy problem, for they know about their economic problems anyway. They want specific help with leprosy. All we can offer is:

1 Early detection and treatment of cases. Efforts should be concentrated on detecting smear-positive cases. There is no agreement on whether or not a leprosy control programme should be fully integrated with the normal health services, or should be a separate project.

2 BCG (bacillus Calmette–Guérin) vaccination. There is conflicting evidence on the effectiveness of BCG in protecting against leprosy. An effectiveness of 80% was reported from Uganda, but elsewhere results have usually been less impressive.

3 Health education and an open treatment policy. It is generally agreed that imprisoning leprosy cases in leprosaria to isolate them from the community is unproductive. The reason lies in the repercussions of the policy: patients will not come forward and tend to be concealed by their relatives. They remain in the community and remain infective because they do not receive specific treatment. So transmission of infection is actually encouraged.

Effect of drugs on infectivity

Lepromatous cases usually become non-infective after 6 months' treatment with dapsone alone. Rifampicin probably makes patients non-infective within 2 weeks.

The WHO is of the opinion that if 75% of all cases were treated for

75% of the time, transmission of the disease would be arrested. This has yet to be proved.

HOPES FOR THE FUTURE

The main hope is for an effective vaccine. The armadillo is useful, but the failure to grow the bacillus in culture is a big handicap. Although dapsone is cheap and usually effective, its action in sterilizing infective cases is slow, and it may induce resistance. There is still a need for new drugs as bactericidal as rifampicin, and some of the new quinolones and tetracyclines show promise; but cost may limit their usefulness.

Health education, if effective, could greatly reduce the numbers of patients presenting with irreversible neurological complications. Early diagnosis and treatment help both the individual and the community.

QUESTIONS, PROBLEMS AND CASES

Consult the text if the answer is not given in the appendix.

1 How do you carry out a skin-slit smear examination in leprosy?

2 Make a table showing the main differences between lepromatous and tuberculoid leprosy. (Use the information on pp. 294–7.)

3 Describe the drug treatment of a patient with lepromatous leprosy.

4 What is erythema nodosum leprosum (ENL)? How would you treat a patient with lepromatous leprosy and severe ENL? (See pp. 303–4.)

5 How may leprosy endanger the eyes?

6 In what types of leprosy can the nerves be involved?

7 What type of leprosy typically causes thickening of one cutaneous nerve?

8 In what sort(s) of leprosy is nerve involvement often bilaterally symmetrical?

9 An African patient complains of a pale but otherwise asymptomatic skin lesion. What clinical findings would make you suspect leprosy?

10 What diseases cause thickened peripheral nerves?

11 A 15-year-old Asian schoolboy with a nodular skin eruption on his face is sent to your clinic by his schoolmaster. The schoolmaster suspects leprosy.

(a) List all the things you would do to reach a correct diagnosis.
(b) Is the lepromin test likely to be helpful in such cases?

FURTHER READING

Bryceson, A., Pfaltzgraff, R.E. (1990) *Leprosy*, 3rd edn. Churchill Livingstone, Edinburgh.
Nordeen S.K. (1992) Elimination of leprosy as a public health problem. *Leprosy Rev* **63**, 1–4.

Tropical skin conditions

CHAPTER 37

Tropical ulcers

Ulcers are among the most common skin lesions in the tropics. Chronic ulcers due to different causes are often difficult to identify, and as the treatment varies greatly with the cause, some procedure must be applied to sort them out. We use the following simple scheme:
• An ulcer of more than 2 months' duration with no distinguishing features: treat on general principles.
• Has the ulcer healed 6 weeks later? No: investigate further. Yes: no further action needed.

The most important exception is the oral or genital ulcer, which may be a syphilitic chancre. But it is not really a chronic ulcer.

Tropical ulcer (TrU) is a chronic ulcer which heals with non-specific treatment. Such ulcers are widespread in tropical countries and may be related to local injury or insect bites. More than 90% of lesions are below the knee. The cause is not known, but Vincent's organisms (a fusiform bacillus and a spirochaete) and β-haemolytic streptococci are often isolated on culture. The single most commonly cultured organisms are *Fusobacterium* spp., anaerobic Gram-negative rods which produce butyric acid. There is no evidence of direct person-to-person transmission. Tropical ulcers are common in tropical Africa, South-east Asia and south America.

CLINICAL PICTURE AND NATURAL HISTORY

The first sign is often a painful papule or blister below the knee. This soon breaks down to form an ulcer, which is painful and which rapidly extends in the next few weeks. The edges of the ulcer are typically slightly raised, swollen, red and tender, and not undermined. The floor of the ulcer, which penetrates the superficial fascia, is composed of granulation tissue and covered in pus. There may be only one ulcer, or several. There are never very many. After a few weeks acute inflammation subsides and the ulcer stops increasing in size. The mature static ulcer is usually 1–10 cm in diameter. Some ulcers heal spontaneously, but many remain open for years.

EFFECTS OF INADEQUATE TREATMENT

Most patients with TrU are treated in dispensaries or similar places. The dispenser will usually clean the ulcer with an antiseptic, apply a gauze dressing soaked with acriflavine or something similar, and then put on a bandage. When the patient comes back a few days later for removal of the dressing, pus has usually soaked through the gauze and dried. The gauze is then firmly adherent to the ulcer, and newly grown skin will have become enmeshed by the gauze. The dresser removes the bandage. To shorten the pain, he or she briskly rips off the dressing. This removes all the new epidermis that has grown since the last visit, the evidence of which is in the circle of bleeding which surrounds the ulcer. The repetition of this procedure eventually convinces even the most optimistic of patients that further attendance at the clinic is useless. The patient then treats him- or herself by washing the ulcer periodically to remove pus and slough. The water is often grossly contaminated, and if a dressing is used, it will certainly not be sterile. After some months or years, the floor of the ulcer becomes raised, due to an increasing amount of fibrous tissue being formed in its base. Spontaneous healing has now become unlikely, and the ulcer may persist until the patient dies.

COMPLICATIONS

The most important are:
1 Infection of underlying bone or joint (osteitis, arthritis).
2 Septicaemia.
3 Epithelioma formation in the edge of the ulcer.
4 Secondary infection with toxigenic diphtheria organisms.
5 Tetanus.

DIAGNOSIS

The clinical picture is sufficiently characteristic for a provisional diagnosis to be made clinically. It is confirmed mainly by:
1 Exclusion of other causes.
2 Satisfactory response to non-specific treatment.
   A therapeutic trial is often very useful. There can be no specific diagnosis of TrU, so purists say the condition does not exist. But the syndrome described above certainly does exist, and it responds to this treatment. What we lack is an aetiology, not a disease. However, the *Bacillus fusiformis* is a promising candidate for the villain.

TREATMENT

This is logical and comprises three stages:

1 Removal of pus and slough and elimination of surface pathogens. The patient should be put to bed if possible, the ulcer should be cleaned with 1% hydrogen peroxide or Eusol or dilute sodium hypochlorite solution (BNF, BPC) and a wet antiseptic dressing, kept wet by being covered with a sheet of polyvinyl chloride or polyethylene, should be applied and renewed daily. At each change of dressing the wound should be cleaned to remove exudate and slough. The ulcer usually appears clean after 3–7 days of this treatment.

2 Elimination of bacteria in the tissues. This is achieved by injections of 0.6–1.2 Mu of procaine or triple penicillin intramuscularly daily for 3–7 days. All the causal organisms respond to penicillin. Pain disappears in 1–3 days.

3 Application of non-adherent dressings to permit healing. Once the ulcer is clean, a non-adherent dressing is applied, and left in place as long as possible to allow the new skin to grow undisturbed, usually 10–14 days. Paraffin gauze (tulle gras) is the usual northern remedy, but the results with equivalent indigenous remedies, such as gauze impregnated with coconut oil in Tanzania, are often just as good and much less expensive. Fancy paraffin gauze containing topical antibiotics produces no better results, and is more likely to cause contact dermatitis. Plain gauze liberally spread with 3% chlortetracycline ointment works well if nothing else is available.

When renewing the dressing, great care should be taken to avoid damaging the delicate skin at the edge of the ulcer. If the ulcer is found to be infected when the dressing is removed, it should be dealt with by using wet antiseptic dressings as described under (1) daily until it is clean. The non-adherent dressing is then resumed.

This procedure, repeated as often as necessary, will result in the healing of all TrUs of recent onset within a few weeks. If any islands of skin are present in the ulcer floor they should be preserved, as they form foci for the regrowth of skin. The end-result of medical treatment is a thin, papery scar. If this overlies bone it may be vulnerable to trauma, and require some sort of protection. We have seen several ex-Far-East prisoners of war whose old TrU scars break down from time to time 40 or more years after the original ulcer healed.

Treatment of chronic TrU with fibrosis

Elevated ulcers with a dense fibrous substrate will often fail to heal with conservative treatment. The only remedy then is to excise the fibrous tissue and apply a skin graft. If the ulcer overlies bone, the best results will be obtained with a cross-leg pedicle graft. Other ulcers will heal with free grafts, but phases (1) and (2) of the conservative regimen must always precede grafting to minimize infection.

EPIDEMIOLOGY AND CONTROL

TrU is a disease of the countryside, commonest in people working on the land. The only way to reduce the incidence of TrU is to ensure, by provision of good primary care, that minor wounds and sores are treated promptly and effectively. All primary health care workers must be taught to apply the treatment method outlined to established or incipient TrU.

There is no good evidence that TrU is related to malnutrition.

CHAPTER 38

Buruli ulcer

Buruli ulcer is a spreading, highly destructive ulcerating condition caused by *Mycobacterium ulcerans*, named after the place in Uganda where it was first identified. It may affect any part of the body, but mainly the limbs. The reservoir and mode of spread are not known, but direct person-to-person spread does not seem to occur. The disease has been reported from several parts of Africa, Papua New Guinea, the Americas and Southeast Asia.

CLINICAL PICTURE

The ulcer is often preceded by a firm, painless nodule which may be itchy. The nodule grows, the overlying skin breaks down, and the ulcer forms and often spreads very rapidly in all directions. The following features are clinically very characteristic:

1 The ulcer is painless.

2 The skin at the edge of the ulcer is deeply undermined.

3 Satellite ulcers often communicate with the original ulcer by a subcutaneous tunnel, so the skin between adjacent ulcers is often unattached to the underlying tissues. The extent of the damage is always much greater than it looks from the surface.

NATURAL HISTORY

After initial rapid extension, the ulcer tends to heal over months or years. There may be extensive scarring; contracted scar tissue often leads to gross deformities. The worst lesions lead to degloving of the skin of the arms, and comparable damage in the leg. There is usually no secondary infection.

DIAGNOSIS

Acid-fast bacilli may be found in the base of the ulcer and in its edges. The bacilli grow at 33°C on media suitable for *M. tuberculosis*, but colonies may take up to 14 weeks to appear.

TREATMENT

Radical excision and debridement followed by skin grafting were used initially. It has since been shown that much of the undermined skin can be conserved, with a great shortening of healing time and reduction of scarring, if the skin can be incised to form flaps so that both surfaces can be cleaned and treated by local antiseptic applications. Silver nitrate (0.5%) is one of several antiseptics found to be effective. Afterwards the skin flaps are replaced and held in position by sterile dressings and a crêpe bandage. Clofazimine is the most active drug against *M. ulcerans*, given in a dose of 100 mg daily until the ulcer has healed. Rifampicin also has useful activity. The other antituberculous drugs do not. Skin grafting is needed in all but the smallest ulcers. Doctors working in endemic areas may acquire enough experience to recognize the preulcerative nodule. If the nodule is recognized and adequately excised, the development of an ulcer can be prevented.

EPIDEMIOLOGY AND CONTROL

As the mode of spread is not understood, nothing can yet be done to prevent it. The infection is often confined to a sharply defined locality, suggesting a reservoir of the infection in nature, but the only isolates so far have been made from foliage.

Cutaneous leishmaniasis

Cutaneous leishmaniasis (CL) is among the most important causes of chronic ulcerating skin lesions in the world. The organisms in the host tissues are mainly found in reticuloendothelial cells in the skin, where as amastigotes they multiply by simple fission. Microscopically they cannot be distinguished from *Leishmania donovani* (Chapter 2). The geographic distribution of cutaneous leishmaniasis is shown in Fig. 39.1.

The parasite has a life cycle virtually identical to that described for *L. donovani*, and the vectors are always female sandflies of the genera *Phlebotomus* or *Lutzomyia*. Amastigotes present in skin lesions are taken up by the flies, and cyclical development takes place in the same way as in *L. donovani* infection (Chapter 2). Cutaneous and mucocutaneous leishmaniases are usually zoonoses, but in some of them the animal reservoir has not been identified.

CLASSIFICATION

The subject has always been confusing. The trouble is that parasites that look identical under the microscope may cause entirely different clinical diseases, and be adapted to different animal hosts. In the past few years understanding of the problem has been greatly advanced by the discovery that the various isoenzymes of different groups of parasites seem to provide a stable indicator of their identity. The importance of this has not been so much in improving the treatment of individual cases, but in understanding the epidemiology of the infection. For example, in a certain part of Africa it may be found that humans and a local rodent seem to be infected with cutaneous leishmaniasis. The natural inference would be that elimination of the rodents would eliminate human disease. But analysis might show that isolates of organisms from humans and the rodent have different isoenzymes. This means that the human infection is not the same as the rodent infection, and that to attack the rodent reservoir would be a complete waste of resources.

As with blood grouping, isoenzymes can prove that two strains are different, but they cannot prove that two strains that appear the same are identical.

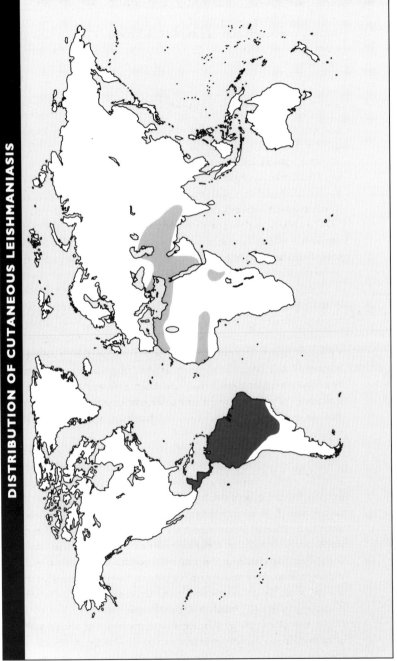

Fig. 39.1 Distribution of cutaneous leishmaniasis. ■ New World; ▨ Old World, *L. tropica*, *L. major* and *L. aethiopica*.

CLINICAL SPECTRUM

Several syndromes can be recognized clinically (Table 39.1). They do not always harmonize with the taxonomic systems described, but medicine is like that. Most patients with CL can be fitted into one of the following categories.

1 Non-ulcerating ('dry') often single lesions

A single, dry, cutaneous lesion, usually with a tendency to remain solitary, to heal without complications over many months, and confer lifelong immunity against infection with the same parasite. This is the typical picture of urban CL in the Middle East (*L. tropica*). This is the most benign of the CL syndromes and, but for facial scarring in young women, it would scarcely deserve attention. The reservoir is probably humans only. Infections in dogs are probably a reverse zoonosis in which the animal is infected from humans. Nelson's terminology of zoo-anthro-ponosis is used mainly by his friends, and has not yet become widely current.

2 Multiple, often exudative ('wet') lesions

This is typical of *L. major* (sometimes incorrectly called *L. tropica*, var. *major*) and several central and south American parasites. The greater severity of the lesions may reflect their shorter evolutionary association with humans, as all are normally zoonoses. Before ulceration occurs, there is usually a pronounced cellular response to the parasites, taking the form of a densely infiltrated papule which may reach 2 cm or more in diameter.

3 Relapsing lesions

Relapses following apparent healing of lesions occur in several south and central American parasites. We have seen this picture several times in British soldiers from Belize, infected with *L. mexicana*. Relapses usually only occur two or three times before final, spontaneous cure. It is only because chicleros (the men who go into the forests to collect the chicle latex from which chewing gum is made) are bitten so often and in such disparate parts of the body that they develop the distinctive destructive lesion of the pinna called a chiclero's ulcer. It is probably the result of a destructive primary infection rather than a result of metastasis.

CL SYNDROMES

Species	Main distribution	Main diseases	Reservoir
OLD WORLD			
L. tropica	North-west India, Pakistan, Afghanistan, Mediterranean area, west Asia	Oriental sore, urban CL, recidivans CL	Humans (dog probably an incidental host)
L. major	Former USSR east of the Caspian, Afghanistan, Iran, Arabia, north Africa, Sudan, Senegal, Niger	Oriental sore, rural CL (multiple sores are frequent)	Various desert rodents
L. aethiopica	Ethiopia, Mount Elgon (Kenya)	Oriental sore, small number of cases develop disseminated CL (DCL)	Hyraxes
NEW WORLD			
L. mexicana complex			
L. mexicana	Belize, other parts of central America	South American CL, chiclero's ulcer, DCL	Forest rodents
L. amazonensis	Brazil	South American CL, DCL	Forest rodents
L. braziliensis complex			
L. braziliensis	Much of south America as far north as Belize	South American CL, espundia, mucocutaneous leishmaniasis	Unknown— possibly dog, horse
L. guyanensis	Guyanas and adjacent Venezuela and Brazil	South American CL, pian bois	Sloths and other forest mammals
L. panamensis	Panama, probably other parts of south America	South American CL (very rarely mucocutaneous disease)	Sloths
L. peruviana	Peruvian Andes	South American CL (Uta)	? Dog

Table 39.1 The main cutaneous leishmaniasis (CL) syndromes.

4 Chronic relapsing lesions

Some CL lesions become very chronic and persist for years, often waxing and waning in severity, and sometimes almost disappearing. This is the picture of recidivans CL, and occurs in a small proportion of patients with *L. tropica* infection. It is almost certainly a result of a hypersensitivity to leishmanial antigen on the part of the host. Scarring is a prominent feature in late cases.

5 Diffuse lesions

Wide dissemination of CL gives the picture of disseminated CL or DCL. Diffuse CL is a synonym for the condition. Its presence denotes deficient host CMI, but it is most often associated with *L. aethiopica* and *L. amazonensis*.

6 Mucocutaneous lesions

These are the most destructive lesions of all, probably greatly enhanced by secondary infection with aerobic and anaerobic bacteria. The lesions usually involve the mucocutaneous junction of the nose, and spread inwards to destroy the adjacent tissues over a period of years. The parasite most likely to cause this is *L. braziliensis*, but other organisms such as *L. panamensis* may also be responsible at times. All these lesions are preceded by cutaneous lesions indistinguishable from those described in (1) and (2).

IMMUNITY AND CUTANEOUS LEISHMANIASIS

Destruction of the amastigotes responsible for the lesions of CL is mainly by CMI. The same parasite inoculated into different hosts of the same species may produce clinical lesions covering the spectrum from a small, solitary, self-limiting ulcer to diffuse infiltration of the skin with nodule formation, much as in lepromatous leprosy. The analogy between CL and leprosy is very close. This certainly extends to antibody production, which tends to be greatest in cases where CMI is least effective. High levels of antibody are usually found in DCL.

TYPICAL CL ULCER

The ulcer may be single, multiple, and on any part of the body, but usually is on skin not covered by clothes. The edge is typically raised, infiltrated and not undermined. Most lesions are relatively painless. The regional lymph glands may or may not be enlarged. But some ulcers are apparently very superficial, and show little evidence of marginal infiltration. The ulcers are most often found on the extremities and on the face, and there is a marked tendency to spontaneous cure.

RECIDIVANS PICTURE

This is an ulcer which waxes and wanes in time with the balance of the pathogenicity of the parasite and the effectiveness of the host's defences. This biological battle may continue for many years. The battle leaves scars, and the combination of scarring and signs of active inflammation is characteristic of the condition.

PICTURE OF DISSEMINATED CUTANEOUS LEISHMANIASIS

Diffuse or DCL closely resembles lepromatous leprosy. There is diffuse or nodular infiltration of the skin with enormous numbers of amastigotes in phagocytic cells. Ulceration is unusual, probably because of deficient CMI. There may be extensive depigmentation in the areas of affected skin, increasing the resemblance to leprosy.

MUCOCUTANEOUS LEISHMANIASIS

The main feature that distinguishes this from other sorts of CL is the partiality of the parasites for the mucocutaneous junctions of the nose. The first perceptible lesion is often a nodule adjacent to the nostril. Granulomatous, destructive lesions with chronic ulceration follow. After many years, the nasal septum and the other nasal cartilaginous structures may be destroyed, replacing the nose with a grotesque cavity in the centre of the face. The role of secondary infection is arguable, but may be very important. This is the only form of non-visceral leishmaniasis which carries a significant mortality.

DIAGNOSIS

Direct diagnosis

Only direct diagnosis is completely dependable. Material must be removed from the active edge of a lesion using a wide-bore needle, or by the skin-slit smear technique as used in leprosy (Chapter 36, pp. 298–9). Smears from the base of an ulcer are useless. The material is spread on a slide, dried, fixed and stained with Giemsa or a similar Romanowsky stain. The amastigotes exactly resemble those of kala-azar, and most of them will be found inside macrophages. Material inoculated into special culture medium (such as NNN) or into hamsters may be essential for the diagnosis of recidivans CL, in which microscopy alone is often negative. Cultures are also necessary to provide enough material for isoenzyme studies. Biopsy may be needed in the more difficult cases, but impression smears should be made before fixing the specimen. Giemsa-stained sections are best. The amastigotes always look much smaller in sections than in smears.

Indirect diagnosis

A variety of serological tests will detect antibody after CL has been present for some weeks, such as the IFAT using promastigotes in culture as antigen. The main use of serodiagnostic methods is when CL is suspected clinically but direct diagnostic methods have failed. In endemic areas a high proportion of the population may be seropositive. Unfortunately, there seems to be some cross-reaction with lepromatous leprosy, so the distinction between DCL and LL leprosy cannot be made reliably by serology. Fortunately, bacilli are always easy to find in LL leprosy. The leishmanin test (Montenegro test) detects a delayed sensitivity reaction to killed promastigotes injected intradermally. It is read at 48 h, like the tuberculin test. It is often negative during the acute stages of CL and becomes positive after cure. In endemic areas a high proportion of the population will be leishmanin test-positive and may also have healed scars. It is negative in DCL. A strongly positive test may be useful in diagnosing recidivans cases, because the routine histology from these lesions is often indistinguishable from *Lupus vulgaris*.

TREATMENT

Many forms of CL are self-limiting and require no treatment other than protection from secondary infection. This particularly applies to *Leishmania tropica* and *L. mexicana*. When treatment is needed, systemic pentavalent antimony (Chapter 2, p. 45) is usually effective. Lesions can also be infiltrated with pentavalent antimony. Response varies with the different parasites and also from person to person, presumably due to variations in immunity. Suspect mucocutaneous leishmaniasis requires prolonged systemic pentavalent antimony. Metronidazole seems to have useful activity in espundia, probably due to its effect on anaerobic secondary infection. Early cases usually respond to a pentavalent antimony drug. Diffuse CL responds poorly to pentavalent antimony but has been treated with pentamidine. Other drugs effective against visceral leishmaniasis (Chapter 2, pp. 45–6) may be tried in resistant cases. Simple curettage will cure many early CL lesions.

EPIDEMIOLOGY AND CONTROL

Except in *L. tropica*, where humans are the reservoir of infection and the disease is often urban, these are zoonotic infections and control of the wild animal reservoir is largely impossible. In towns where *L. tropica* is endemic, sandfly control is usually impracticable. Because lifelong immunity usually follows CL due to *L. tropica*, uninfected children are often deliberately inoculated with material from an active lesion, using a needle. The preferred sites are the arm or leg. This has been carried out for generations in parts of the Middle East, because CL lesions on the face can cause a cosmetic blemish in girls, and many naturally acquired lesions occur on the face. From this potentially dangerous practice, with its risk of transmitting hepatitis B and other infections, has developed the widespread use of vaccines made from cultures, especially in Israel, Jordan and the former USSR. Vaccines for the other forms of CL are still fairly recent and experimental. The practical difficulties of using a fragile live vaccine are obvious. The cross-immunity between the various types of CL is not fully understood, but *L. major* vaccine has been used to protect against *L. major* and *L. tropica* infections.

In areas where CL is due to *L. tropica*, dogs are often found to be infected, with a similar or possibly identical parasite. But the dog is not now believed to be the reservoir of infection. Some degree of protection against the forest zoonotic CL disease of south America can be given to inhabitants of new settlements by separating the houses from the forest by a wide zone cleared of trees. However, people going into the forest to hunt and to gather wood cannot yet be protected. Rodent control by the Russians in Soviet Asia has been very successful in eliminating *L. major* in settlers.

FURTHER READING

Molyneux, D.H., Ashford, R.W. (1983) *The Biology of* Trypanosoma *and* Leishmania *Parasites in Man and Domestic Animals.* Taylor & Francis, London.

Peters, W., Killick-Kendrick, R. (eds) (1987) *The Leishmaniases in Biology and Medicine.* Academic Press, London.

WHO (1990) *Control of the Leishmaniases.* Technical Report Series 793. World Health Organization, Geneva.

Ulcers in general

Ulcers that heal on non-specific treatment will include minor injuries with secondary infection and tropical ulcers. Some of the ulcers with a specific aetiology will also heal initially, but will break down later. Some forms of cutaneous leishmaniasis often do this.

Table 40.1 lists the most important chronic ulcers.

CHRONIC ULCERS

Type of ulcer	Main characteristics	How diagnosis is established
Tropical ulcer	Painful; usually lower leg	Heals on non-specific regimen and does not recur in same place
Buruli ulcer	Extensive; mainly painless; very deep undermined edges	Finding acid-fast bacilli in ulcer
Cutaneous leishmaniasis	Single or multiple; often with infiltrated edges; not undermined	Amastigotes in edges of lesions
Desert sore, veld sore (cutaneous diphtheria)	Usually single, painful onset with vesicle; adherent slough; undermined; paralysis from toxin	Culture. The combination of sore + slough + palsy is diagnostic
Tertiary syphilis (gumma)	Chronic, usually painless ulcers on extremity	Serological tests for syphilis. Spirochaetes cannot be found
Tuberculous ulcer	Frank ulcers often follow subcutaneous TB; there may be adjacent cold abscess; evidence of TB elsewhere	Microscopy for AFB. Culture
Sickle-cell disease in adults	Rare in Africa where few adults with the disease survive. Ulcers often symmetrical on lower legs	Patients is obviously anaemic; sickling test is positive; more than 90% of the Hb is HbS on electrophoresis
Dracontiasis (the guinea worm)	The pearly prolapsed uterus is seen early; later the worm itself	By identifying the worm, or larvae, expelled after exposure to water
Trophic ulcer of leprosy	Painless; may be deeply penetrating on the sole	Associated evidence of nerve damage: loss of sensation

Table 40.1 Causes of chronic ulcers in the tropics.

Itchy skin lesions in the tropics

The main causes of itchy skin lesions in the tropics are:

1. Scabies.
2. Superficial fungus infections.
3. Eczema, which may be due to topical sensitization or drugs.
4. Onchocerciasis.
5. African trypanosomiasis.
6. Cutaneous larva migrans.
7. Larva currens due to *Strongyloides stercoralis* infection.
8. Insect bites.

ITCH MITE

The commonest cause of itch is probably the itch mite, *Sarcoptes scabei*. There is usually a 6-week period between infection and the onset of symptoms. In small children the lesions are often very widespread and secondarily infected, so the classical interdigital pattern may be completely obscured. A papular or pustular itchy skin eruption involving the buttocks is particularly suspect, and justifies a therapeutic trial of benzyl benzoate even if burrows cannot be found. Scabies is not entirely benign, for secondary infection of the excoriated lesions with nephritogenic strains of *Streptococcus* may cause acute nephritis.

Scabies is a household infection: if one member has symptomatic infection, the other members of the household must be treated willy-nilly. Prevention of scabies can be achieved by the routine use of soap containing 5% tetraethylthiuram monosulphide (Tetmosol). In some countries this is no more expensive than ordinary toilet soap.

CUTANEOUS LARVA MIGRANS (CREEPING ERUPTION)

This is most often caused by the intradermal migration of *Ancylostoma braziliense*, the common tropical and subtropical hookworm of dogs. The disease affects anyone coming into contact with soil contaminated with the faeces of dogs harbouring the parasite. The life cycle is like that of a human hookworm, except that when humans are infected by penetration of the skin by the filariform larvae, the larvae develop no further, and wander around in the skin for many weeks before they finally die. Their migration causes an intensely itchy skin eruption—a red, excoriated track which accurately records the aimless wanderings of the parasite.

Most lesions occur on the feet, but small children infected by playing in contaminated sandpits or on beaches often have lesions on the buttocks and genitalia. The track usually advances only a few millimetres a day, in contrast to the very speedy movements of *Strongyloides stercoralis* larvae in larva currens (Chapter 20, p. 198).

Cutaneous larva migrans due to *A. braziliense* can usually be effectively treated by local freezing of the head of the advancing track, using an ethyl chloride spray, or by the application of a home-made ointment containing 10% metriphonate or 10% thiabendazole in an emulsifying base applied for 12 h under an occlusive dressing. When the patient has multiple lesions, a course of albendazole (400 mg twice a day for 5 days) is dramatically effective. Our last such patient said 'after 48 hours, doctor, they all stopped moving'.

Miscellaneous exotic conditions

There are many miscellaneous exotic conditions if one includes the large number of insect larvae that occasionally invade human skin. Only two are common: myiasis due to *Cordylobia anthropophaga* (the tumbu or putsi fly) and chiggers due to *Tunga penetrans*.

TUMBU FLY

Tumbu fly infection is acquired when the larvae of *C. anthropophaga* penetrate human skin, and is widely distributed in west and central Africa. The normal cycle is between the fly and various rodents. Eggs laid on clothing rapidly develop into invasive larvae when exposed to body warmth, and the larvae penetrate the skin where they mature in about 2 weeks. If undisturbed, the larvae then fall to the ground, pupate and develop into mature flies in a few days.

Tumbu fly infection presents as multiple 'blind boils'. The head of the 'boil' has black marks corresponding to the spiracles of the larval fly.

Treatment

The easiest way is to suffocate the larva, and remove it intact when it shows signs of distress. The methods are all variations on the same theme: the spiracles must be obstructed with an oily substance. Old engine oil, petroleum jelly, liquid paraffin and glycerine and ichthyol are all effective.

The application of the oil is followed by vigorous wriggling movements of the larva, and when it partly emerges, it can be grasped with forceps and removed intact. Incomplete removal of a damaged larva causes a fierce inflammatory reaction, so undue force to pull it out must not be used.

CHIGGERS

Chiggers are the lesions produced when the gravid female flea, *T. penetrans*, establishes herself in a human host. The usual site is a toe.

The pig is the natural host. The infection is acquired by walking barefoot in endemic areas, which include the Americas, east, west and southern Africa and the west coast of India. The condition is diagnosed by the appearance of a painful, localized swelling at the side of a toenail, and the demonstration of the bloated, egg-filled body of the female flea on unroofing the lesion. Treatment is by microsurgery.

The lesion is exposed by dissection with a needle, and the female flea is removed intact. In endemic areas, great skill in this procedure is often possessed by the local people, but sepsis or tetanus may ensue. An antiseptic dressing should be applied when the flea has been removed, to prevent secondary infection.

Snake bite

Snake bite

INTRODUCTION

The medically important venomous snakes all have fangs at the front of the upper jaw and may be classified as follows:

Vipers

Land snakes with long erectile fangs and a distinct division between head and neck. The venom usually (but not invariably) contains haemorrhagic and procoagulant components causing bleeding and incoagulable blood. Local effects may include swelling and local necrosis.

Elapids

Land snakes, including cobras, mamba, kraits, coral snakes and all Australasian snakes. They have short fixed front fangs with no distinct division between head and neck. The venom usually (but not invariably) contains neurotoxic components which can cause respiratory and less severe paralysis. Local effects are variable.

Sea snakes

These have small heads and long, thin bodies and prominent flattened tails. They rarely bite. The venom contains neurotoxic and myotoxic components.

These notes outline general first-aid principles and the medical management of envenoming. The following general facts should be considered when dealing with victims.

1 Snakes only bite when they feel threatened; only 50–70% of bites by venomous snakes result in significant envenoming.

2 Antivenom should only be given when there is a definite indication that it is needed.

3 The victim should be reassured immediately.

FIRST AID

The aims of first aid should be to treat or delay life-threatening effects which may develop before the patient reaches medical care, to hasten the safe transport of the patient to a medical facility and to avoid harmful effects. Measures should ideally be short, simple, practicable and helpful.

1 Avoid panic.

2 Determine the cause of the accident except where this puts the patient or others at further risk. If the snake has already been killed, it should be brought to the hospital with the patient, but take care not to handle the dead animal directly.

3 Lie the patient on his or her side to maintain a clear airway and to prevent inhalation of vomit.

4 Wounds should be wiped with a clean cloth to remove superficial venom.

5 Avoid extreme measures such as wound incision, suction, tourniquets, electric shock therapy, ice packs, local application or injection of chemicals or the use of any proprietary kits and potentially dangerous traditional methods. On admission to hospital, remove any tourniquet that has been applied.

6 Do not give the patient anything by mouth. Water should only be given to avoid dehydration in the event of delay in receiving medical treatment.

7 For bites by snakes which are known not to cause local necrosis and swelling (e.g. Australasian elapids, kraits, sea snakes, coral snakes, etc.) the use of a pressure immobilization bandage and splint is recommended. The bandage should be applied from the bite site over the entire length of the bitten limb; it should be firm as for a sprain and peripheral pulses should be readily detectable. There is some evidence that this will delay the absorption of venom into the circulation.

8 Introduction of venom into the eye by the spitting cobras should be treated by liberal irrigation with water or other available fluids (e.g. milk, urine, etc.).

MEDICAL TREATMENT

1 The diagnosis, the presence and severity of envenoming and the timing and evolution of the symptoms should be established. The time from the accident, the site of the bite and the snake responsible should be determined as far as is possible and the patient treated accordingly.

2 Clinical evaluation of envenoming (e.g. local changes, systemic bleeding, coagulation status, hypovolaemic shock, vasodilation or myocardial damage, neuromuscular paralysis, rhabdomyolysis, renal failure and tender enlargement of lymph glands draining the area of the injury) should be carried out.

3 Laboratory evaluation of envenoming should be performed. The extent of the tests used depends on the level of sophistication of the laboratory.

4 Patients with no signs of envenoming on admission should still be observed for up to 24 h. If no signs develop within this period they may be discharged.

5 Criteria for systemic envenoming which require immediate treatment with antivenom include the following:

(a) Coagulopathy (incoagulable blood), haemorrhage (spontaneous bleeding from gums, old wounds, etc.) occurs mainly, but not invariably, in viper bite. These signs will only be reversed following neutralization of all circulating venom by antivenom. Antivenom dose can be assessed by the resolution of coagulopathy using the simple whole-blood clotting test (2 ml of blood is placed in a new, clean dry glass test tube which is left upright and undisturbed at room temperature for 20 min. The tube is then tipped gently to see if the blood is still liquid and runs out freely (non-clotting) or if a clot has formed. If the blood fails to clot 6 h after the first or subsequent doses of antivenom, a further dose should be administered. Replacement therapy (e.g. fresh frozen plasma, etc.) is not generally useful and should only be considered if no antivenom is available.

(b) Neurotoxic paralysis occurs mainly, but not invariably, in elapid and sea snake bite (e.g. ptosis, ophthalmoplegia, dysarthria, peripheral muscle weakness, respiratory distress). Early antivenom treatment (e.g. on observation of ptosis) may prevent progression to complete respiratory paralysis. If paralysis occurs, prolonged ventilatory support (hours or days) may be required. Anticholinesterase treatment may be useful in reversing postsynaptic neuromuscular blockade (e.g. most cobra venoms) if antivenom is not available. If there is response to a test dose of edrophonium (Tensilon), continuation of treatment with neostigmine may be given, as in myasthenia gravis. Presynaptic neuromuscular blockade usually responds poorly to antivenom.

(c) Rhabdomyolysis (tender muscles, trismus, hyperkalaemia, myoglobinuria). Venom myolysins usually affect skeletal muscle only. Dark urine, elevated serum creatine phosphokinase and transaminases are common findings and in some severe cases (e.g. bites by some sea snakes and the south American tropical rattlesnake, *Crotalus durissus terrificus*) there may be severe hyperkalaemia and acute renal failure. Muscle weakness caused by venom myolysins may be mistaken for neuromuscular paralysis.

(d) Cardiotoxicity (arrhythmias, heart failure).

(e) Impairment of renal function occurs sometimes in viperine envenoming. Renal failure may produce a high mortality in some areas (e.g. Russell's viper, *Daboia russelli*, in south-east Asia). Antivenom does not appear to be effective in reversing this effect and catheterization and careful fluid balance monitoring is advisable. Dialysis may be required in severe cases.

6 Local tissue injury (e.g. necrosis) should be treated as for a burn. Antibiotics are required for specific infection. In the rare event of a compartment syndrome, fasciotomy may be considered, but only after correction by antivenom of any venom-induced coagulopathy or haemorrhage. Local signs such as necrosis or swelling do not respond well to antivenom and, alone, they are not an indication for antivenom. However, antivenom is recommended if the swelling extends to more than half the bitten limb. In some viper bites, swelling may be extensive, resulting in massive extravasation of fluid into the bitten limb with rapid development of hypovolaemic and toxic shock. Aggressive fluid replacement, initially with crystalloids, should be carried out in such cases.

7 Appropriate antibiotics should be used if infection is present. Tetanus prophylaxis should be administered routinely.

ANTIVENOM

Antivenom administration

Antivenom is made by hyperimmunizing large animals with either the venom from one species of snake (monospecific) or with a mixture of venoms from a number of species (polyspecific). The extent of purification varies, ranging from a relatively pure F(ab)$_2$ fraction of the IgG molecule to filtered crude serum or plasma. Antivenom is therefore potentially hazardous and should be used only if specifically indicated for treatment of envenoming. The only currently recommended route for antivenom is intravenously, either by drip over 30 min or by slow 'push' injection. Children should receive the same dose of antivenom as an adult. During and immediately after antivenom, the patient should be carefully monitored for evidence of anaphylaxis. In the event of a reaction, antivenom should be stopped and subcutaneous adrenaline (0.5 ml 1:1000) and parenteral antihistamine given; once the reaction has subsided the antivenom should be restarted and continued until the full dose has been administered. If the patient has a history of allergy, there may be a case for pretreatment with adrenaline and antihistamines before antivenom. Skin sensitivity testing before treatment with antivenom is unreliable and not recommended. Serum sickness (e.g. arthralgia, urticaria, etc.) may occur 5–10 days after antivenom therapy and should be treated with steroids and antihistamines, depending on the severity of the reaction.

The main reasons for the failure of antivenom treatment are incorrect antivenom, insufficient dose, wrong route and opaque material. The expiry date on the ampoule or vial is not important, providing the antivenom remains clear, potency is not effected. A good general rule is that it is never too late to give antivenom.

List of available antivenoms

Registers of commercially available antivenoms are essential reference material for those responsible for treating bites and stings by venomous animals, for advising ministries of health, hospitals, international health agencies and expeditions. The most recent information may be obtained from the list compiled by Theakston and Warrell (1991, *Toxicon*, 29, 1419–70).

FURTHER READING

Maegraith, B.G. (1989) Snake bite and other venomous bites and stings (contribution by R.D.G. Theakston). In: *Adams and Maegraith: Clinical Tropical Diseases*, 9th ed. Blackwell Scientific Publications, Oxford, pp. 332–45.

Reid, H.A., Theakston, R.D.G. (1983) The management of snake bite. *Bull WHO* 61, 885–95.

Warrell, D.A. (1983) Injuries, envenoming, poisoning, and allergic reactions caused by animals. Section 6, vol. I. In: Weatherall, D.J., Ledingham, J.G.G., Warrell, D.A. (eds) *Oxford Textbook of Medicine*. Oxford University Press, Oxford, pp. 6.35–42.

Answers

I (a) By the bite of an infected anopheline mosquito.
(b) By blood transfusion.
(c) Transplacentally.
(d) By inoculation (needlestick injury).

2 Because of the various species of human malaria, only *P. falciparum* causes severe disease, with system or organ complications that may be fatal.

3 (a) By haemolysis (of both parasitized and unparasitized red cells).
(b) From secondary hypersplenism.
(c) By temporary marrow suppression.
(d) Occasionally by haemorrhage when there is severe DIC.

4 (a) Malaria.
(b) Gastroenteritis of various aetiologies (viral and bacterial).
(c) Localized and generalized infections of almost any sort, including the exanthemata.

5 The thick blood film, stained with Romanowsky stain at pH 7.2. But in holoendemic areas, the finding of a few parasites in a thick blood film does not necessarily mean that malaria is causing the current illness, because so many residents have asymptomatic parasitaemia. A high parasitaemia is more likely to be significant than a low one in such cases.

6 (a) Check the blood glucose concentration; if low, give 50% dextrose (1 ml/kg body weight) intravenously.
(b) Stop the convulsions (with diazepam by injection).
(c) Reduce the temperature (if above 39°C) by cooling and by giving rectal paracetamol.
(d) Examine the child and take a blood film. If positive, treat with parenteral quinine. If the blood film is negative, repeat blood film after 4 and 8 h.
(e) Whether the blood film is positive or not, do a lumbar puncture. If the result suggests pyogenic meningitis, treat accordingly.

7 They should all use permethrin-impregated mosquito nets at night. They all need drug prophylaxis. Probably the best option now (1995) is mefloquine 250 mg, one tablet weekly for the adults and one-quarter tablet weekly for the children. Recommendations change with time.

8 It can be very difficult, and in many cases impossible. The fever caused by malaria can certainly cause convulsions in children prone to febrile convulsions. Usually consciousness is regained within a few minutes (less than half an hour) after a febrile convulsion, while a child with cerebral malaria who has a convulsion will remain comatose for much longer. The distinction is largely academic, because in both cases one would do the same: give an anticonvulsant and treat the malaria.

9 (a) Typhoid fever. Scanty malarial parasitaemia is of no significance in an adult brought up in a holoendemic area.

(b) Blood and stool cultures.

(c) Chloramphenicol or ciprofloxacin (for 14 days). He should also be given an antimalarial drug, e.g. Fansidar 3 tablets.

10 (a) A thick blood film. This showed numerous trophozoites and a few schizonts of *P. falciparum*. On the thin film, the parasitaemia was about 10%.

(b) Yes. In cerebral malaria there may be focal CNS signs in addition to evidence of diffuse cerebral disturbance.

(c) A lumber puncture. This showed a protein level of 60 mg/dl and a raised WBC of 20/µl, all lymphocytes. The glucose was normal.

(d) Yes, but the CSF in cerebral malaria is often entirely normal.

(e) He needs quinine by slow intravenous infusion. The fits must be controlled. Nursing efforts should be directed towards preventing aspiration and, monitoring fluid intake and output, blood pressure and temperature.

(f) No! Dexamethasone prolongs coma, increases the complication rate and does not improve survival. As soon as the patient can swallow, change to oral medication.

(g) No. Even when evidence of DIC is present in malaria, heparin should not be given. Controlled trial has shown that heparin may be lethal in such cases, often causing death from cerebral haemorrhage. This is presumably related to pre-existing capillary damage. But fresh plasma and platelet concentrates are safe, and may help.

CHAPTER 2

1 (a) Dogs.
   (b) Wild canines, such as jackals.
   (c) Rodents.
   (d) Humans.

2 (a) Pneumonia due to pneumococcus (*Streptococcus pneumoniae*).
   (b) Dysentery due to various species of *Shigella* (bacillary dysentery).
   (c) Dysentery due to *Entamoeba histolytica* (amoebic dysentery).
   (d) Tuberculosis.

3 (a) No.
   (b) Yes.
   (c) Partly.
   (d) Yes, due to hypersplenism (mainly).
   (e) Not usually.

4 (a) Purpura.
   (b) Bruises.
   (c) External bleeding (e.g. epistaxis).

5 (a) Raised globulins (IgG; polyclonal).
   (b) Reduced production of immunoglobulins protective against other infections.
   (c) Diminished albumin synthesis.

6 (a) Increased susceptibility to infections.
   (b) Oedema.

7 (a) A blood film for parasites.
   (b) A full blood count and differential WBC.
   (c) A chest X-ray.
Sadly blood culture, essential for diagnosing brucellosis, is often not available, but serology may be helpful.

Comments:

(a) The obvious differential tropical diagnoses are malaria and visceral leishmaniasis. I thought he was a classic case of kala-azar, knowing that it occurred in the area he had been living in. In fact his blood film showed parasites of *P malariae*, both trophozoites and schizonts. He had not been taking antimalarial chemoprophylaxis.

(b) The important results needed are the haemoglobin level (or haematocrit), and an opinion on the appearances of the blood film. Thrombocytopenia should be apparent from the film, if it is severe. A red cell count is a waste of time unless an automatic cell counting machine is available.

(c) The chest X-ray may reveal localized or disseminated (miliary) changes. Pulmonary tuberculosis is an important complication of visceral leishmaniasis, and should be suspected in any patient who fails to respond promptly to specific antileishmaniasis chemotherapy.

8 (a) A test for leishmanial antibodies. This was strongly positive (using promastigotes in culture, and the IFAT was positive at a titre of 1 in 256).

(b) A smear from the buffy coat of centrifuged blood. (No amastigotes were found in the Giemsa smear.)

(c) Blood culture on NNN medium (this was negative).

(d) Marrow smear and culture. This is the most important test. Even with a low platelet count, this investigation should be safe and should have been done even in the presence of a positive immunological test. If no amastigotes were found in the biopsy, two other possible diagnoses—aleukaemic leukaemia and disseminated tuberculosis— might well be brought to light.

(e) A chest X-ray (which might show a primary tuberculous lesion or miliary shadowing). A Mantoux test would not be of much help, as it would probably be negative whatever the diagnosis, even in miliary tuberculosis.

Outcome: She responded dramatically to a course of Pentostam. We still use her serum as a control positive for the *Leishmania* IFAT.

CHAPTER 3

1 (a) *T. rhodesiense* is less well-adapted to humans, so causes a more severe and acute general illness.

(b) In *T. gambiense* the patient usually dies in the late stage due to slowly progressive CNS disease.

(c) In *T. rhodesiense* death often occurs in the acute toxaemic stage from myocardial involvement.

(d) Serous cavity effusions are common in *T. rhodesiense* infections.

2 None.

3 Gland puncture.

4 A blood film, or a more sensitive test for detecting trypomastigotes in blood.

5 (a) A bone marrow smear.

(b) A smear from the CSF deposit.

6 A greatly raised IgM level.

7 (a) F. The cells would be mainly polymorphs, unless the meningitis had been partly treated, when it could give this picture.

(b) F. *T. rhodesiense* does not extend to west Africa.

(c) T.

(d) Only if he had not been brought up in the area or been on prolonged chemoprophylaxis. An indigenous adult would be expected to be immune to the serious effects of *Plasmodium falciparum* otherwise.

(e) T. Definitely yes.

(f) T.

(g) T.

8 (a) T. A common feature in late disease.

(b) F (except over the chancre).

(c) T. These are very characteristic.

(d) T. But only visible in fair-skinned people.

9 (a) F. Both can cause infections resembling the other in an individual case: *T. gambiense* may be atypically acute. *T. rhodesiense* uncharacteristically indolent.

(b) T. Yes: *T. rhodesiense* is rapidly lethal in white rats, *T. gambiense* is not.

(c) F. An early hope, but disappointed.

(d) F. There is apparently complete cross-reaction between antibodies raised to each parasite. There is more hope of differentiation by isoenzyme identification.

10 It is too toxic. A patient in the early stages of the disease may die from arsenical encephalopathy due to malarsoprol, when a much less toxic drug could equally well have achieved a cure.

11 (a) By the bite of a tsetse fly infective by reason of cylical development of the parasite within its body. The saliva contains the infective metacyclic trypanosomes.

(b) Mechanically, by the proboscis of a biting fly (mainly the tsetse).

(c) Congenitally by transplacental transmission. ('Blood transfusion' would also count as a correct answer.)

12 The possibility of infecting the CSF by allowing the ingress of trypanosomes should be prevented by a prior injection of suramin.

13 The most effective measure would involve active case-finding and treatment. The first essential would be to meet the village headman, explain the situation, and obtain his cooperation. After this short-term measure, you would try to identify places of human–fly contact and persuade the local people to implement the measures outlined previously (see Principles of control, p. 63) so as to minimize this danger. Special tsetse-fly traps or impregnated screens may greatly reduce the fly population.

14 (a) T.
  (b) T (typical of *T. rhodesiense*).
  (c) T.
  (d) T.

CHAPTER 12

1 (a) (i) ALA.
  (ii) Pyogenic liver abscess.
  (iii) Hepatoma.
  (iv) Active (viral) hepatitis.
  (v) Congestive cardiac failure, perhaps associated with subacute bacterial endocarditis.
  (vi) An infected hydatid cyst.
  (b) (i) Liver scan (isotope or ultrasound).
  (ii) Amoebic antibody measurement.
  (iii) WBC and differential.
  (iv) α-Fetoprotein measurement (typically raised in hepatoma).

2 No. If he has dysentery he will be excreting amoebae, which are not infective. Only amoebic cysts are infective.

3 (a) Ileocaecal TB (much more common than amoeboma).
  (b) Appendix abscess (rare in some parts of the world such as rural Africa, relatively common in others such as urban South-east Asia).
  (c) Salpingitis or pyosalpinx (usually gonococcal).
  (d) Ruptured ectopic gestation (very common where c is common, as tubal pregnancy often follows tubal damage).
  (e) Amoeboma.
  (f) Deep mycosis, such as ileocaecal actinomycosis.
  (g) Crohn's disease (rare in the tropics).
  (h) Yersinial ileitis (frequency in tropics unknown).

Comment: Carcinoma would be rare in any part of the world in this age group. In most parts of the tropics large bowel carcinoma is uncommon. This situation may change with the wider adoption of a European diet.

4 (a) No. Amoebic infection alone would not explain the high fever and other severe constitutional symptoms.

(b) Stool culture. A heavy growth of *Campylobacter jejuni* was obtained on special medium under microaerophilic conditions. This doubtless accounted for his initial symptoms. A *Shigella* infection could equally well have done so.

(c) Specific antiamoebic therapy. His *Campylobacter* infection is not causing his current symptoms and will be eliminated spontaneously.

5 Non-specific hepatitis accompanying colitis of any cause. Amoebic hepatitis, a diffuse inflammation of the liver caused by amoebae in the absence of abscess formation, is a largely discredited entity. Focal infiltration of the liver with inflammatory cells may accompany colitis of any aetiology: bacterial, amoebic or non-specific. It subsides spontaneously when the colitis itself recovers. It is presumably a result of microembolization of the liver with bacteria entering the portal system via the damaged colonic epithelium.

6 Treat him for ALA despite the clinical picture resembling hepatoma more closely. In the event, he made a complete recovery on oral metronidazole. No patient with a clinical diagnosis of hepatoma should be allowed to die without receiving antiamoebic treatment, because the clinical differentiation of the two conditions is often impossible.

CHAPTER 21

1 (a) Amoebic dysentery.
(b) Heavy *Trichuris* infection.
(c) *Schistosoma mansoni* or *S. japonicum* infection.
(d) Balantidial dysentery.

2 (a) Are there dogs in the house?
(b) Does she play in an area contaminated by dogs?
(c) Does she eat dirt?
(d) *Toxocara* antibody tests if possible.

3 (a) (i) *Toxoplasma gondii* infection.
(ii) *Toxocara* infection.
(iii) Cysticercosis and several other rarer larval helminths.
(b) (i) The Sabin–Feldman dye test.
(ii) *Toxocara* antibody test.
(iii) Cysticercal serology.

4 *Ascaris lumbricoides*, or an earthworm eaten in salad (a spurious infection). The earthworm will have a ringed surface.

5 (a) Amoebic liver abscess: fever, leucocytosis, high ESR, anaemia, positive serology, liver ultrasound.

(b) Pyogenic liver abscess: fever, leucocytosis, high ESR, negative amoebic serology, liver ultrasound.

(c) *Ascaris* abscess: small child, heavy *Ascaris* infection (high egg count), jaundice.

(d) Secondarily infected hydatid cyst.

(e) Very early stage of viral hepatitis.

6 (a) Strongyloidiasis.

(b) Finding typical larvae in the stools. Less common causes include *Capillaria phillipenensis*, and very heavy infection with an intestinal fluke such as *Fasciolopsis buskii* (in the Far East).

7 (a) *Strongyloides stercoralis* infection—look for larvae in the stools and sputum.

(b) Amoebiasis—find amoebic trophozoites in the fresh stool.

8 (a) Trypanosomiasis.

(b) Strongyloidiasis.

9 (a) Death with diarrhoea and CNS depression.

(b) Desferrioxamine (Desferal).

10 (a) Continuing blood loss from another source such as a duodenal ulcer.

(b) Undetected concomitant folate deficiency.

(c) Undetected infection such as pulmonary TB.

(d) A haemoglobinopathy such as thalassaemia or HbSC disease.

(e) Uraemia.

11 (a) The finding of a hypochromic anaemia, with low mean corpuscular haemoglobin concentration (MCHC) and hypochromic cells on the blood film. The mean corpuscular volume (MCV) and MCHC can only be accurately determined if an automatic red-cell counting machine is available. They would both be low.

(b) The finding of numerous hookworm eggs in the faecal smear.

(c) The absence of any other obvious cause of blood loss or features suggesting malabsorption.

CHAPTER 25

1 (a) The incubation period is the time elapsing between cercarial penetration and the development of symptoms. In asymptomatic infections it is therefore infinitely long. In heavy infections causing the Katayama syndrome it may be as short as 4 weeks.

(b) The prepatent period is the time elapsing between cercarial penetration and egg passage. It may be as short as 4–6 weeks in *S. mansoni* infections, but is more commonly 3 months or more.

2 (a) Visitors to endemic areas.
(b) Those exposed to an initial heavy infection.
(c) People first exposed to *S. mansoni* and *S. japonicum* infection.

3 An invasive helminth infection.
(a) *S. mansoni* and *S. japonicum*.
(b) *Strongyloides stercoralis* (upper abdominal pain is common).
(c) *Fasciola hepatica* (there may be associated anaemia due to haemobilia).
(d) *Capillaria philippinensis* infection.

4 (a) Urinary schistosomiasis and acute glomerulonephritis, which may be associated with streptococcal skin sepsis, such as secondarily infected scabies.
(b) Is the first part of the stream relatively free of blood, and most blood passed towards the end of micturition? If 'yes', this indicates terminal haematuria, typical of urinary schistosomiasis. Dysuria in the form of pain on micturition referred to the tip of the penis would also favour schistosomiasis.
(c) Examination of the centrifuged deposit from a urine sample. Discrete red cells would be recognized in either case, but in schistosomiasis the typical terminal-spined eggs should be easily found. In glomerulonephritis, there will be numerous casts, mainly granular.
(d) No. And it would subject the patient to the needless risk of secondary infection. Cystoscopy is indicated in a patient with a past history of urinary schistosomiasis who develops haematuria later in life: this is often the first symptom of a bladder carcinoma.
(e) No. Direct diagnosis by the finding of eggs in the urine is the best and only unequivocal diagnostic method. The only reservation to remember is that dead eggs may continue to be expelled in the urine of a patient for years after all the adult worms are dead. Dead eggs will not hatch. They are often opaque.

5 Treat him for the schistosomiasis and repeat the pyelogram 3 months later. If the obstructive uropathy was due to granulomatous tissue, it will have improved or resolved. If it was due to fibrous ureteric stricture, there will be no improvement. Surgery is only needed if monitoring shows deterioration in renal function.

6 Examination of the urine for eggs. Viable eggs were present in large numbers. This could not have been due to re-exposure, as he had been living in the UK for 1 year before he was first seen and had remained here ever since. The conclusion: he had either not taken his medicine (metriphonate in this case) or the drug had been ineffective. His uncle said he had not taken the medicine. A repeat course under personal supervision resulted in parasitological cure and regression of the obstructive lesions.

7 (a) Site workers' houses as far away from irrigation canals and drainage ditches as possible.

(b) Provide a piped water supply.

(c) Install individual latrines for each household (communal latrines quickly become dirty, and are then not used).

(d) Provide a safe pool with frequent changes of water for children to play in, in the workers' village.

(e) Ensure all the workers are examined for schistosomiasis and treated appropriately.

(f) Ensure all new workers arriving after the scheme starts operating are also examined for schistosomiasis and treated as necessary. A high turnover rate may make this very important—but expensive.

(g) Ensure annual re-examination of all people living on the estate to monitor the degree of transmission.

(h) Apply molluscicides (a job for experts) if transmission can be identified as occurring in the irrigation system.

In one such estate, staff and their families were becoming very heavily infected despite most of the precautions, but the infection was acquired outside the estate. The correct identification of transmission sites is essential before starting an expensive molluscciding programme.

CHAPTER 29

1 (a) Tuberculous lymph glands in the groin.

(b) Damage to the lymph glands following *Chlamydia granulomatis* infection (lymphogranuloma venereum).

(c) Bancroftian filariasis.

(d) Kaposi's sarcoma of the classical (not HIV-related) type. (A relatively common tumour in Africans. It can cause changes very like elephantiasis.)

(e) Siderosilicosis of the inguinal lymph glands.

(f) Recurrent erysipelas.

The history could be very helpful, and knowledge of the common local conditions. For example, in a large tecaching hospital in west Africa, much the commonest cause of this syndrome was found to be

TB. But in parts of east Africa and in Asia, filariasis might be far more common. In certain parts of the tropics where people walk barefoot on iron-rich, silicate soils, particles of silicate trapped in the inguinal glands set up a fibrosing condition which causes lymphatic obstruction. This condition, which is known as podoconiosis, has been most reported from Ethiopia and Papua New Guinea.

2 (a)   (i) Do you feel otherwise well? Is there any fever?

(ii) Are there any other painful areas, or have you noticed any sores?

(iii) Have you had any recent sexual contact?

(iv) Has anyone you know of had anything like it recently?

(b)   (i) The entire area drained by the glands should be carefully searched for a related lesion. This includes not only the leg, foot and external genitalia, but also the perineum and anal region.

(ii) Evidence of lymphangitis.

(iii) Elevation of temperature.

(c)   (i) Pyogenic infection from a focus in the same lymphatic territory, most often the foot. The red streaks of lymphangitis might be visible.

(ii) A venereal infection if associated with a previous or present genital sore, or one near the anal margin. Most likely would be lymphogranuloma venereum or chancroid, but syphilitic adenopathy is not always painless.

(iii) An acute episode in *W. bancrofti* or *Brugia malayi* infection.

(iv) Plague.

(v) Rat-bite fever due to *Spirillum minus*. There may be an associated rash.

(vi) Scrub typhus.

CHAPTER 34

Treat him with a full course of DEC, and hope for the best.

CHAPTER 35

1 (a)   (i) Have you noticed blood in the stool? (His answer was no.)

(ii) Have you been having abdominal pain? ('Yes, in the upper abdomen').

(iii) Have you noticed any change in the colour of the stool? ('Yes, it has been black at times and after a motion I have sometimes felt faint').

(b) A gastrointestinal lesion causing bleeding.

(c) Not really. Abdominal pain is seldom a feature of hookworm infection and neither is recurrent melaena.

(d) Stool microscopy for ova and parasites. Stool examination for occult blood. A barium meal examination.

The occult blood test was strongly positive; he had five hookworm eggs per coverslip on direct smear examination, and the radiologist reported a large duodenal ulcer.

Comment: Duodenal ulcers are quite common in parts of the rural tropics. Repeated small bleeds may cause the iron-deficiency picture described. One large bleed would cause a normochromic, normocytic anaemia with brisk reticulocytosis.

2 (a) A megaloblastic anaemia.

    (b) (i) Obstetric history (grand multiparity favours folate deficiency).

        (ii) Diet (a strict vegetarian diet favours B_{12} deficiency).

        (iii) Stools (steatorrhoea due to tropical sprue is an important cause of folate deficiency).

Comment: She had only had three pregnancies, but her diet, apart from a very small amount of milk, seemed to be entirely devoid of animal products. The diagnosis of nutritional B_{12} deficiency was confirmed by her prompt response to oral hydroxocobalamin. If she had had Addisonian pernicious anaemia (rare in most parts of the tropics), she would not have responded to B_{12} by mouth.

3 (a) Haemolysis due to the malaria. The urine will probably contain an excess of urobilinogen, and the serum indirect bilirubin will probably be raised. Although usually more severe in primiparas, malaria can affect any pregnancy.

(b) A course of chloroquine.

(c) Regular malaria prophylaxis and daily folic acid.

4 (a) β-Thalassaemia major.

(b) The development of haemosiderosis will be accelerated.

(c) Very poor. Even if blood transfusion is freely available, death from cardiac haemosiderosis early in the third decade is usual despite the use of chelating agents. They are only really effective if given by infusion pump.

5 (a) Folate deficiency associated with tropical sprue. Investigations showed a low serum folate level, greatly increased fat excretion, and the jejunal biopsy revealed some flattening of the villi with submucosal infiltration. If the history had been longer, he might have had a low B_{12} level too. Patients with tropical sprue cannot absorb vitamin B_{12}, but blood levels do not fall until the stores are used up.

(b) Stool microscopy for parasites (especially *Giardia* and *Strongyloides*) and a barium meal. His alcohol intake might be relevant. The stool showed fat droplets only and the barium meal no evidence of lymphoma or other focal pathology.

(c) Folic acid by mouth for at least 3 months and tetracycline 250 mg four times daily for 3 weeks. Tetracycline may need to be continued up to 3 months in some cases.

6 (a) Sickle-cell disease.

(b) A normochromic, normocytic anaemia. Sickled cells may be seen in ordinary blood films during a crisis, but between crises may require incubating with a reducing substance to reveal them. The RI is usually raised.

(c) In malarious areas, continuous antimalarial prophylaxis. Daily folic acid supplements, as their need for folate is increased.

(d) (i) A megaloblastic crisis due to folate deficiency.

(ii) A hyperhaemolytic crisis, maybe due to malaria.

(iii) An aplastic crisis due to an intercurrent parvovirus infection.

(iv) A sequestration crisis with rapid enlargement of liver and/or spleen.

7 Sickle-cell Hb C (HbSC) disease—a much less serious condition than sickle-cell anaemia (HbSS) and fully compatible with a normal life and life expectancy, except in pregnancy.

Females with HbSS in the tropics seldom achieve reproductive competence. Women with HbSC often do, and pregnancy may result in a dangerously severe haemolytic anaemia. Blood transfusion with frusemide may be life-saving.

8 The mild anaemia of non-descript type fits a diagnosis of secondary anaemia due to infection. This is supported by the fever, weight loss and raised ESR. The urinary findings are typical of renal TB. Further investigations were the Heaf test (4+), intravenous pyelogram (an abnormality of the left kidney consistent with renal TB) and, finally, the finding of AFB in a centrifuged deposit of the early-morning urine.

Specific antituberculous chemotherapy was followed by a rapid clinical response.

CHAPTER 36

Answers to questions 1–4 are given in the text.

5 (a) Acute iridocyclitis, in LL disease and in ENL reactions.

(b) By involvement of the facial nerve, causing paralysis of the orbicularis oculi muscle, and inability to protect the eye by blinking (lagophthalmos).

(c) By causing anaesthesia of the cornea, involving the first division of the fifth nerve.

6 In all types of leprosy. Nerve involvement is usually asymmetrical in tuberculoid disease, symmetrical in lepromatous disease. Nerves are often involved very severely in the borderline types, especially during reactions.

7 TT or BT disease.

8 LL disease.

9 Loss of sensation, sweating or hair loss. Thickening of a nearby nerve, or a palpable nerve in the lesion.

10 (a) Leprosy.
(b) Amyloidosis.
(c) Familial hypertrophic neuropathy.

11 (a) (i) Examine the patient carefully. Look especially for infiltration of the tissues, involvement of the ears and nasal mucosa, changes in pigmentation, loss of eyebrows, thickening of nerve trunks and peripheral nerve function. Do the nodules have a comedone in the centre? If they do, acne is the most likely cause.
(ii) Take a careful family history to discover if any close contacts have a similar condition.
(iii) Perform a skin-slit smear examination on one of the nodules. Prepare at least two slides. Stain one with modified Ziehl–Neelsen stain for *M. leprae,* the other with Giemsa stain to detect amastigotes of leishmaniasis. The patient could be suffering from post-kala-azar dermal leishmaniasis. In Ethiopia it could be disseminated cutaneous leishmaniasis.
(b) No. It will be negative if the patient has lepromatous leprosy. It could also be negative in any other condition. Although a positive test would exclude lepromatous leprosy, the skin-slit smear is a quicker way of doing so.

Index

Abdominal tuberculosis 106
Abscess
  amoebic liver 133, 145–6
    aspiration 152–3
    diagnosis 149–50
    treatment 152–3
  typhoid 97
Aedes aegypti mosquito
  dengue fever 78
  dengue haemorrhagic fever 81–2
  yellow fever 79
African tick typhus 75–6
African trypanosomiasis see
        Trypanosomiasis, African
AIDS 122
  tuberculosis 102, 107
  visceral leishmaniasis 42
  see also Human immunodeficiency
        virus (HIV)
AIDS-related complex 121
Albendazole
  ascariasis 184
  cutaneous larva migrans 329
  giardiasis 159
  hookworm infection 192
  hydatid disease 217
  strongyloidiasis 199, 201
  Taenia solium infection 213
  toxocariasis 203
  trichuriasis 195
Allopurinol, visceral leishmaniasis 46
Amastigotes, Leishmania donovani 38
American tick typhus 76
Amodiaquine, malaria 21, 29
Amoebiasis 141–55
  abdominal symptoms, 'vague' 144
  asymptomatic, treatment 153
  clinical picture 142–7
  control 154
  cutaneous 146–7
  diagnosis 148
  drug prophylaxis 154
  epidemiology 153–4
  fulminating 144

parasite and life cycle 141–2
postdysenteric colitis 145
treatment 150–3
For Questions, problems and cases
    see pp. 154–5
Amoebic dysentery
  bacillary dysentery,
        differentiation 143
  clinical picture 142–4
  diagnosis 147
  pathogenesis 142
  treatment 151–2
  untreated, natural history 143–4
Amoebic liver abscess 133, 145–6
  aspiration 152–3
  diagnosis 149–50
  treatment 152–3
Amoebic pus 145
Amoebic ulcer 142, 144
Amoebicides 150–1
Amoeboma 144–5
Amoxycillin
  diarrhoea 172
  typhoid fever 99, 100
Amphotericin B, visceral
        leishmaniasis 46
Ampicillin
  diarrhoea 172
  typhoid fever 99, 100
Anaemia
  amoebic liver abscess 150
  categories 279–82
  diagnostic process 283–8
  fever 134
  folate deficiency 280, 284, 287–8
  haematological parameters 282
  haemolytic
    malaria 10
    typhoid fever 97
  hypochromic 283, 284, 286
  investigations 281–2
  iron-deficiency 280, 285–6
    hookworm infection 186–93
  macrocytic 284, 287

Anaemia (contd)
    megaloblastic 284, 287–8
    normochromic 283, 284, 286
    normochromic normocytic 286
    normocytic 283, 286–7
    physical signs 280
    vitamin B$_{12}$ deficiency 280, 283–4,
        287–8
    For Questions, problems and cases
        see pp. 288–90
Anaesthesia of extremities,
        leprosy 305–6
Ancylostoma braziliense 328–9, Plate
        20
Ancylostoma duodenale 185, 186, 189,
        192
    see also Hookworm
Anopheles spp., Malayan filariasis 258
Antigenic variation 54
Antivenom 338–9
Arbovirus infections 77–84
    encephalitis 82–3
    fever (with or without rash) 78–9,
        81
    haemorrhagic fever syndrome
        79–82
Argentinian haemorrhagic fever 85–6
Aromatic diamidine drugs, visceral
        leishmaniasis 45–6
Ascaris lumbricoides infection 192
    diagnosis 183
    effects of adult worms 182
    epidemiology and control 184
    life cycle 181
    presentation 182
    surgery 182
    symptoms due to pulmonary
        migration 181
    treatment 183–4
Ascaris suum 181
Ascites, chronic fever 137–8
Ascorbic acid, hookworm
        anaemia 191

Bacillary dysentery 171, 172
    amoebic dysentery,
        differentiation 143
Bacillus thuringiensis 270
Balantidiasis 161
Balantidium coli 161, 174
BCG vaccine, leprosy 307

Benzyl benzoate, scabies 328
Benzylpenicillin, HIV 124
Bephenium hydroxynaphthoate,
        hookworm infection 192
Bilharziasis see Schistosomiasis
Blackwater fever 13
Bladder calcification,
        schistosomiasis 226–7
Blindness, onchocerciasis 265
Bolivian haemorrhagic fever 85–6
Bone tuberculosis 106
Borrelia spp., relapsing fever 70
Brugia malayi infection 258–9, 275
Brugia pahangi infection 275
Burkitt's lymphoma 16
Buruli ulcer 317–18, 327

Calabar swellings 260–1
Campylobacter infection 170, 171,
        172–3
Capillaria philippinensis infection 234
Capsaicin, diarrhoea 177
Card agglutination test for
        trypanosomiasis (CATT) 58
Cardiomegaly, in South American
        trypanosomiasis 67–8
Casoni test, hydatid disease 217
Cell-mediated immunity (CMI)
    Cutaneous leishmaniasis 323
    leprosy 297–8
    visceral leishmaniasis 40, 41
Cercopithecus aethiops 86
Cerebral toxoplasmosis, HIV
        infection 128
Chagas disease 67–9
Chagoma 67
Chiclero's ulcer 321
Chiggers 330–1
Chikungunya viruses 79, 81
Chloramphenicol
    African tick typhus 76
    HIV infection 124
    scrub typhus 75
    typhoid fever 99
Chloroquine
    amoebiasis 151
    malaria 19–21
        prophylaxis 5, 28–9
        resistance 22–3
            classification 24–5
Chlorpromazine, cholera 167

Chlortetracycline
    cholera 167
    tropical ulcer (ointment) 315
Cholecystitis, typhoid fever 97, 100
Cholera
    changes in fluid and
        electrolytes 163
    classical disease 162–3
    clinical spectrum 163
    control 167–8
    diagnosis 163–4
    epidemiology 167
    maintenance hydration 165–6
    treatment 164–7
        rehydration 164–5
Chrysops flies, loiasis 260, 262
Cinchonism 22
Ciprofloxacin, diarrhoea 172
Circumoval granuloma,
        schistosomiasis 224–5
Clioquinol, amoebiasis 151, 154
Clofazimine
    Buruli ulcer 318
    erythema nodosum leprosum 304,
        Plate 18
    leprosy 301
Clonorchis sinensis 240–1
Clostridial food poisoning 176–7
Clostridium difficile 175
Clostridium perfringens 170, 176
Colitis, postdysenteric 145
Colon, pseudopolyposis,
        schistosomiasis 227–8
Complement activation, African
        trypanosomiasis 54
Complement fixation test (CFT)
    hydatid disease 217
    South American
        trypanosomiasis 68
Cor pulmonale, schistosomiasis 227
Cordylobia anthropophaga 330
Corticosteroids see Steroids
Co-trimoxazole
    diarrhoea 172
    HIV infection 124
    typhoid fever 100
Cough, HIV infection
    acute 123–4
    chronic 124–5
Coumarin, elephantiasis 256
Countercurrent
        immunoelectrophoresis (CIE),

    hydatid disease 217
Crotalus durissus 337
Cryptosporidium 178
Culex mosquitos
    Rift Valley fever 84
    viral encephalitis 82
Culex quinquefasciatus 256, 257
Cutaneous amoebiasis 146–7
Cyclops, Dracunculus medinensis
        infection 271, 274
Cystericercosis, human 211, 212
    cerebral 212
    treatment 213
Cysticercus bovis 209
Cysticercus cellulosae 211

Daboia russelli 338
Dapsone
    leprosy 301–2, 307
        resistance 302, Plate 8
    malaria 27, 29
DDT
    malaria 33–4
    visceral leishmaniasis 48
Dehydroemetine
    amoebiasis 151
    amoebic dysentery
Dengue fever 78, 81
Dengue haemorrhagic fever 81–2
Dengue shock syndrome 81–2
Desert sore 327
Diarrhoea
    acute 171–7
        with fever and blood 171–3
        with fever but no blood 173–4
        without fever but with
            blood 174–5
        without fever or blood 175–7
    chronic, HIV infection 126–7
    classification 170–1
    food toxicants 177
    history 169–70
    persistent 177–8
    Strongyloides stercoralis
        infection 198, 199
    travellers' 176
    viral causes 177
Diethylcarbamazine (DEC)
    Brugia malayi infection 259
    loiasis 262
    onchocerciasis 268, Plates 21 & 27

Diethylcarbamazine (contd)
    tropical eosinophilia syndrome 276
    Wuchereria bancrofti infection
        255–6, 257
Difluoromethyl ornithine (DFMO),
    African trypanosomiasis 61
Di-iodohydroxy-quinoline,
    amoebiasis 151
Diloxanide furoate
    amoebiasis 151
        asymptomatic 153
    amoebic dysentery 152
Dimethyl phthalate, scrub typhus 75
Diphtheria, cutaneous 327
Direct agglutination test (DAT),
    visceral leishmaniasis 43
Dirofilaria immitis 255
    complement fixation test, tropical
        eosinophilia syndrome 275
Disseminated intravascular
    coagulation, cerebral
    malaria 13
Dot immunoassay tests, HIV 115
Doxycycline
    malaria 5, 29
    scrub typhus 475
Dracontiasis see Dracunculus
    medinensis infection
Dracunculus medinensis infection
    diagnosis 272
    epidemiology and control 273–4
    parasite and life cycle 271
    pathogenesis 271
    symptoms 271–2
    treatment 272–3
    ulcer (differential diagnosis) 327
Dysentery see Amoebic dysentery
    and Bacillary dysentery

Ebola virus 86–7
Echinococcus infections see Hydatid
    disease
Echinostoma spp. 247
El Tor cholera 162, 167
Elapids, bite 335
    treatment 337
Elephantiasis
    bancroftian filariasis 252–3
        treatment 256
    onchocerciasis 264

Emetine, amoebiasis 151
Emetine bismuth iodide,
    amoebiasis 151
Emetine hydrochloride, amoebic
    dysentery 151
Encephalitis
    caused by arbovirus infection 82–3
    HIV infection 128
Entamoeba hartmanni 142, 153
Entamoeba histolytica 141–2, 144,
        153–4, 177
    cysts 148–9
Enteric fevers see Paratyphoid fever;
        Typhoid fever
Enteroviruses 177
Enzyme-linked immunosorbent assay
    (ELISA)
    HIV infection 114–15
    hydatid disease 217
    strongyloidiasis 200
    visceral leishmaniasis 43
Eosinophilia
    hydatid disease 217
    loiasis 261
    onchocerciasis 268
    oriental liver flukes 241
    schistosomiasis 234
    strongyloidiasis 200–1
    tropical eosinophilia
        syndrome 275–6
Epilepsy, cysticercosis 212
Epstein–Barr virus, malaria 16
Erisipelas de la costa 266
Erythema nodosum leprosum 303–4,
        Plates 11 & 18
Erythromycin, diarrhoea 173
Escherichia coli, diarrhoea 171–2, 176,
        178
Ethambutol, tuberculosis 107, 108
    adverse effects 111
    HIV infection 125
Ethionamide, leprosy 301
Eyes
    leprosy 295, 306
    onchocerciasis 265

Fansidar, malaria 22–3, 29
Fansimef 23
Fasciola hepatica 242
Fasciolopsis buski 246–7

Fernandez reaction 300
Ferrous sulphate, hookworm
        anaemia 190–1
Fevers
  acute 130
    non-localizing
        accompaniments 132
      with anaemia 134
      with haemorrhagic rash 133–4
      with negative malarial blood
          film 131
      with polymorphonuclear
          leucocytosis and localizing
          symptoms 131
      with polymorphonuclear
          leucocytosis and no obvious
          localizing symptoms 131–2
      with tender liver 132–3
      without polymorphonuclear
          leucocytosis and with negative
          blood film 132
  chronic 130, 134–8
    ascites 137
    causes 134–5
    HIV infection 137
    undetermined origin 135–6
    with normal WBC and
        ESR 136–7
    with relapsing pattern 135
  HIV infection 123–6
  malaria 9–10
  routine investigations 130
Filarial lymphangitis 252
Filoviruses 86–7
Fisherman's itch 222
Flukes
  blood see Schistosomiasis
  intestinal 246–7
  liver, oriental
    clinical picture 241
    diagnosis 242
    epidemiology and control 242
    life cycle of parasites 240–1
    parasites 240
    treatment 242
  lung
    diagnosis 245
    ectopic 244
    epidemiology and control 245
    life cycle 243–4
    treatment 245

Folinic acid, folate deficiency
        anaemia 288
Food toxicants, diarrhoea 177
Footdrop, leprosy 306
Formol gel test 44
Furazolidone
  cholera 167
  giardiasis 159
Fusobacterium spp., tropical ulcer
        313

Gametocytes in malaria 9
Gastrointestinal tuberculosis 106
Genitourinary tuberculosis 106
Giardia lamblia 156, 157, 177, 178
Giardiasis
  clinical picture 156–7
  diagnosis 158
  epidemiology and control 160
  natural history 157
  parasite and life cycle 156
  pathogenesis 157
  treatment 159
Glucose-electrolyte solution,
        cholera 166
Glucose-6-phosphate dehydrogenase
        (G6PD) deficiency
  blackwater fever 13
  malaria 15
Guillain–Barré syndrome, typhoid
        fever 97
Guinea worm infection 271–4, 327
Gumma 327

Haemorrhage, typhoid fever 97
Haemorrhagic fever 79–82
Haemorrhagic rash with fever 133–4
Haemorrhagic viruses 84–7
Hairy string test (Enterotest)
  giardiasis 158
  strongyloidiasis 200
Halofantrine, malaria 5, 23
Hand muscles, leprosy 306
Harada–Mori technique
  hookworm infection 189
  strongyloidiasis 200
Hepatitis
  B 133
  tuberculosis 111

Hepatomegaly
 anaemia 280
 hydatid disease 164
 oriental liver fluke infection 241
 relapsing fever 71
 typhoid fever 96
Herpes zoster, HIV infection 123
Heterophyes heterophyes 247
Hookworm
 anaemia 187–8
  diagnosing 189–90
  factors governing
   development 186–7
  treatment 190–3
 anatomy 185
 clinical effects 186
 culturing eggs 189
 diagnosis 188
 distribution 185
 epidemiology and control 193
 false 188
 life cycle 185–6
Human immunodeficiency virus
  (HIV) 174, 178
 acute seroconversion illness 121
 antiviral therapy 128
 clinical problems 122–8
 control strategies 118
 differences between viruses
   113–14
 disease mechanisms 118–20
 early disease 121
 epidemiology 115–16
 fever 137
 latent infection 121
 natural history of infection 120–2
 paediatric disease 128
 pathogenesis of infection 118–19
 progressive
   immunosuppression 119–20
 prophylaxis 128–9
 risk factors 117
 seroprevalence 116
 staging disease 120
 surveillance 115–16
 testing for 114–15
 transmission 116–17
 see also AIDS
Hydatid disease (*Echinococcus*
  infections) 215–18
 clinical picture 216
 diagnosis 217

epidemiology and control 218
life cycle
  E. granulosus 215–16
  E. multilocularis 215, 216
 treatment 217–18
Hydroxocobalamin, vitamin B_{12}
   deficiency anaemia 288
Hydroxystilbamidine isethionate,
   visceral leishmaniasis 45–6
Hypnozoites in malaria 7
Hypoglycaemia, malaria 13

Immunity
 African trypanosomiasis 54
 HIV infection 119–20
 leishmaniasis
  cutaneous 323
  visceral 41
 leprosy 297, 298
 malaria 14–16
 schistosomiasis 230
 see also Cell-mediated immunity;
   Vaccines
Indirect fluorescent antibody test
   (IFAT)
 amoebic dysentery 147
 leishmaniasis, visceral 43
 malaria 18
 South American
   trypanosomiasis 68
Infection, hypochromic anaemia
   286
Iron deficiency see Anaemia, iron-
   deficiency
Iron dextran, hookworm
   anaemia 191
Isoniazid, tuberculosis 108, 109
 adverse effects 110
Isospora 178
Itch mite 328
Itchy skin lesions 328–9
Ivermectin
 loiasis 262
  onchocerciasis 268–9
  Wuchereria bancrofti infection 256

Japanese encephalitis 82–3
Jaundice 10
Joint tuberculosis 106
Junin virus 86

Kala-azar see Leishmaniasis, visceral
Kaposi's sarcoma 127–8
Katayama fever 224
Kerandel's sign 56
Kidneys, schistosomiasis 229–30
Koilonychia 187, 280
Kyasanur forest fever 79, 81

Lactose intolerance, diarrhoea 178
Larva currens, strongyloides stercoralis
    infection 198, Plate 19
Larva migrans, cutaneous 328–9,
    Plate 20
Lassa fever 84–5
Latex particle agglutination test,
    amoebic liver abscess 149
Leishmania aethiopica 320, 322, 323
Leishmania amazonensis 322, 323
Leishmania braziliensis 322
Leishmania chagasi 38
Leishmania donovani 38, 319, Plate 3
Leishmania guyanensis 322
Leishmania infantum 38
Leishmania major 320, 321, 322, 326
Leishmania mexicana 321, 322, 325,
    Plate 6
Leishmania panamensis 322
Leishmania peruviana 322
Leishmania tropica 320, 322, 325, 326,
    Plates 4 & 5
Leishmaniasis, cutaneous Plates 5 & 6
    classification 319
    clinical spectrum 321–3
    diagnosis 324–5
    disseminated (diffuse) 324
    distribution 320
    epidemiology and control 326
    immunity and 323
    mucocutaneous form 323, 324
    recidivans 324
    treatment 325–6
    ulcers 327
Leishmaniasis, dermal, post-
    kalaazar 44–5
Leishmaniasis, visceral Plate 29
    blood changes 41
    control 48
    diagnosis 42–5
        circumstantial 44
        direct 42–3
        indirect 43

disease, course 40–2
distribution 39
epidemiology 47–8
    patterns of human
        outbreaks 47–8
    established 40–1
    HIV infection 42
    immune changes 41
    life cycle of parasite 38
    natural history 42
    refractory cases 46–7
    treatment 45–7
    For Questions, problems and cases
        see p. 49
Leishmanin test (Montenegro
    test) 40, 41, 325
Leishmanioma 40, Plate 3
Leproma 294
Lepromin test 294, 300
Leprosy Plate 15
    borderline 297
    clinical spectrum 296–7
    control 307
    diagnosis 298–9
    distribution 293
    epidemiology 306–8
    examination 297–8
    histology 300
    indeterminate 294
    initial infection 294
    lepromatous 294–6, Plates 7–8,
        14, 16, 17
    reactions in 303–6
        down-grading 305
        erythema nodosum
            leprosum 303–4
        reversal 304–5, Plates 12 & 13
        upgrading 304–5, Plates 12 & 13
    skin-slit smear 298–9
    transmission 293
    treatment 300–6
        chemotherapy, principles 300–1
        complications 305
        multibacillary leprosy 301–2
        paucibacillary leprosy 302
        persisters 303
    tuberculoid 296, Plates 9 & 10
    ulcers 327
    For Questions, problems and cases
        see p. 308
Levamisole
    ascariasis 184

Levamisole (*contd*)
  hookworm infection 192
Liver
  amoebic liver abscess 150
  anaemia 280
  schistosomiasis 228
  tender, with acute fever 132–3
  see also Flukes
Loa loa infection 260–2, 267
  clinical picture 260–1
  diagnosis 261
  epidemiology 262
  life cycle 260
  prevention 262
  treatment 261–2
Loeffler's syndrome 181
Loiasis 260
Loperamide,
            diarrhoea 176
Louse-borne relapsing fever 70
Lovibond haemoglobinometer 284
Lungs, flukes 244
Lutzomyia spp., cutaneous
            leishmaniasis 319
Lymphadenopathy, tuberculous 105
Lymphangitis, filarial 252
Lymphatic system damage, *Wuchereria
            bancrofti* infection 252–3

Machupo virus 86
Malaria
  blood changes 7–8, 15
  blood films 16–17
    reservations 17
  cerebral 12–13, Plates 1 & 2
  classical stages 10
  clinical features 9–10
  congenital 14
  control, present methods 34–5
  diagnosis 16–18
    new methods 18
  distribution 3–4
  drug resistance 24–5
  epidemiology 30–3
    varieties of patterns 33
  fever 130
  global eradication 33–4
    failure 34
  immune disorders 15–16

immunity 14–16
immunosuppressive effects 16
importance 3
life cycle 3–8
measurement 30–3
morbidity and mortality, measure-
            ment 31
non-immune protective factors 15
pigment 18
pregnancy 14
prepatent period and relapse 7
quartan 8
relapse, problem 25–6
serodiagnosis 17–18
stable 31–2
subtertian 8
tertian 8
treatment 18–24
  chemoprophylaxis 26–30
  chemotherapy, specific 19–24
  drug prophylaxis 5, 35–6
  supportive 18–19
unstable 32–3
untreated, progress 10–11
For Questions, problems and cases
            see pp. 36–7
Malariometric indices 30, 32, 33
Malathion, louse-borne relapsing
            fever 72
Malayan filariasis 258
Maloprim, malaria prophylaxis 29
Mansonella perstans 267
Mansonella streptocerca 267
Mansonia spp., *Brugia malayi* infec-
            tion 258–9
Marburg virus 86–7
Mastomys natalensis 84
Mazzotti test, onchocerciasis 268,
            Plate 27
Mebendazole
  ascariasis 184
  hookworm infection 192
  trichuriasis 195
Mefloquine, malaria 5, 23
  prophylaxis 29
Megacolon, South American
            trypanosomiasis 68
Megaoesophagus, South American
            trypanosomiasis 68
Melarsoprol, African
            trypanosomiasis 60–1

Meningitis
  tuberculous 105–6
  typhoid fever 97
Mepacrine, giardiasis 159
Meptazinol, relapsing fever 72
Merozoites in malaria 3, 9
Metagonimus yokogawai 247
Metriphonate
  cutaneous larva migrans 329
  schistosomiasis 235
Metronidazole
  amoebiasis 151
  amoebic dysentery 151–2
  amoebic liver abscess 152
  cutaneous leishmaniasis 325
  giardiasis 158–9, 178
  guinea worm infection 273
Microfilariae
  Brugia malayi infection, microfilarial
    periodicity 258–9
  Onchocerca volvulus infection 263,
    264
  Wuchereria bancrofti infection
    253–4
    microfilarial periodicity 254–5
Mite typhus 73
Montenegro test (Leishmanin
    test) 40, 41, 325
Mott's morula cells, African
    trypanosomiasis 56
Mycobacterium leprae 293, 300, Plates
    14 & 17
Mycobacterium tuberculosis 102
Mycobacterium ulcerans 317

Nalidixic acid, diarrhoea 172
Nasal smear, leprosy 299
Necator americanus 185, 186, 189, 192
  see also Hookworm
Negri inclusion bodies, rabies 88
Nephrosis, malarial 15–16
Nerves, leprosy 295, 296
Niclosamide
  Taenia saginata treatment 210
  Taenia solium treatment 213
Nifurtimox (Lampit), South American
    trypanosomiasis 68
Nodulectomy, onchocerciasis 269
Norfloxacin, typhoid fever 100

Nose, leprosy 295

Obstructive uropathy,
    schistosomiasis 225–6
Onchocerca volvulus infection Plates
    21–26
  clinical picture 264–5
  diagnosis 266–8
    skin snip technique 267–8
  epidemiology and control 269–70
  parasite and life cycle 263–4
  symptoms 265–6
  treatment 268–9
O'nyongnyong 81
Opisthorchis felineus 240
Opisthorchis viverrini 240–1
Ornithodoros genus, relapsing fever 70
Oxamniquine, schistosomiasis 235
Oxytetracycline
  amoebiasis 151
  cholera 167

Paragonimus africanus 243
Paragonimus westermani 243
Paratyphoid fever 94, 101
  see also Typhoid fever
Paromomycin, amoebiasis 151
Parotitis, typhoid fever 97
Particle agglutination tests, HIV 115
Pediculus humanus, relapsing fever 70
Penicillin
  relapsing fever 72
  tropical ulcer 315
Pentavalent antimony drugs 45, 325
Peritonitis
  tuberculous 137
  typhoid fever 97
Phlebotomus fever 81
Phlebotomus spp, cutaneous
    leishmaniasis 319
Pig bel 177
Piperazine salts, ascariasis 183
Plasmodium falciparum 3, 5, 7, 26,
    170, Plate 1
  anaemia 10
  drug resistance 24, 29–30
  immunity 14–16
  microcirculatory arrest 12

Plasmodium falciparum (contd)
  non-immune protective factors 15
  peculiarities of infection 11–13
  pregnancy 14
  prepatent period 7
  schizogony 8–9
  untreated, progress 10–11
Plasmodium malariae 3, 8
  schizogony 8
  untreated, progress 11
Plasmodium ovale 3, 6, 7
  prepatent period 7
  relapse 25
  schizogony 8
  untreated, progress 11
Plasmodium vivax 3, 5, 6, 7
  congenital malaria 14
  non-immune protective factors
    15
  prepatent period 7
  relapse 25
  schizogony 8
  untreated, progress 11
Pneumonia
  bacterial, HIV infection 123–4
  typhoid fever 97
  verminous 181
Praziquantel
  intestinal flukes 247
  lung flukes 245
  oriental liver flukes 242
  schistosomiasis 236
  Taenia solium infection 213
Prednisolone
  leprosy 305
  tuberculosis 112
Primaquine
  malaria 25–6
  South American
    trypanosomiasis 68
Probenecid, typhoid fever 100
Procaine penicillin, tropical ulcer
  315
Proguanil in malaria 5, 27
  with dapsone 27
Prothionamide, leprosy 301
Pruritus, HIV infection 122–3
Pseudomembranous colitis 175
Pseudopapillomata,
  schistosomiasis 225
Pseudopolyposis of colon,
  schistosomiasis 227

Pulmonary tuberculosis 104–5
Pyrantel embonate
  ascariasis 183
  hookworm infection 192
Pyrazinamide, tuberculosis 108, 109
  adverse effects 111
Pyridoxine deficiency 286
Pyrimethamine, malaria 5, 22–3, 28,
  29

Q fever 76
Qinghaosu, malaria 24
Quinine, malaria 5, 21–2

Rabies
  animals other than dogs 91
  clinical features in man 87–8
  diagnosis 88
  dogs 90–1
  dumb 91
  furious 90
  prevention 91
  treatment 88–90
    WHO guide for post-exposure
      treatment 89, 92–3
Relapsing fever 70–2
Renal disease, typhoid fever 97
Reticulocyte index 281
Ribavirin, lassa fever 85
Ricewater stools, cholera 162
Rickettsia conori var. *pijperi* 75
Rickettsia orientalis 73
Rickettsia rickettsi 76
Rickettsial infections
  African tick typhus 75–6
  American tick typhus 76
  Q fever 76
  scrub typhus 73–5
Rifampicin
  leprosy 301, 302, 305, 307
  tuberculosis 108, 109, 112
    adverse effects 110
Rift Valley fever 79, 83–4
Ringer lactate solution, cholera 165
River blindness 265
Romaña's sign 67
Rotavirus infection 177
Round worm, large *see Ascaris
    lumbricoides*

St Louis encephalitis 83
Salmonella enteritis 173–4
Salmonella infection 100
    schistosomiasis 229
Salmonella paratyphi 94
Salmonella typhi 94
Sandflies, visceral leishmaniasis 47, 48
Sarcoptes scabei 328
Scabies 328
Schistosoma haematobium
        infection 100, 221–2, 224,
        233, 236, Plate 28
    calcification of bladder 226
    cor pulmonale 227
    kidneys 229
    obstructive uropathy 225–6
    rectal biopsy 232
    urine examination in diag-
        nosis 231–2
Schistosoma intercalatum infection 221
Schistosoma japonicum infection 174,
        221–3, 229, 236
    ectopic worms 228
    rectal biopsy 232
    stool examination 232
Schistosoma mansoni infection 100,
        142, 174, 221–3, 235–6
    kidneys 229
    life cycle 223
    liver 228
    pseudopolyposis of colon 227–8
    rectal biopsy 232
    stool examination 232
Schistosomiasis
    control 237–8
    cor pulmonale 227
    diagnosis 231–4
    drug treatment 234–6
        mass 238
    ectopic worms 228–9
    effects of penetration 222–5
    epidemiology 236–8
    human, species affecting 221
    immunity 230
    initial illness 224
    kidneys 229–30
    life cycle of organisms 221–3
    natural history 230–1
    Salmonella infections with 229
    urinary 227
    see also named species
    For Questions, problems and cases

    see pp. 238–9
Schizogony in malaria
    blood 7–8
    site and periodicity 8–9
    tissue 3, 6–7
Schizonticides 19–20
Schizonts of Plasmodium spp. Plate 1
    pre-erythrocytic (PE) 3, 7
Scrub typhus 73–5
Sea snakes, bites 335
    treatment 337
Sexually transmitted diseases (STDs)
        and HIV infection 117
Shigella infection 170, 172
Shingles, HIV 123
Sickle cell disease 15, 174
    ulcers 327
Sickle cell trait 15
Simian immunodeficiency virus
        (SIV) 113
Simulium damnosum 269
Simulium flies, onchocerciasis 263,
        269
Skeletal complications, typhoid
        fever 97
Skin
    changes, onchocerciasis 264
    infections, HIV 123
    leprosy 294–5, 296
    snips, technique for taking 267–8
Skin-slit smear, leprosy 298–9
Sleeping sickness see Trypanosomiasis,
        African
'Slim disease' 126–7
Snake bite 335–9
    antivenom 338–9
    first aid 336
    treatment 336–8
Sodium stibogluconate (Sodium
        antimony gluconate:
        Pentostam), visceral
        leishmaniasis 45
South American trypanosomiasis
        67–9
Spine, tuberculosis 106
Splenectomy, visceral leishmaniasis 47
Splenic puncture, visceral
        leishmaniasis 43
Splenomegaly
    anaemia 280
    hyperreactive malarial 16
    malaria 10, 16

Sporozoites in malaria 3, 9
Staphylococcus aureus, enterotoxin-
    producing strains 175
Steroids
    leprosy 304, 305
    toxocariasis 203
    tuberculosis 112
    typhoid fever 100
Streptomycin, tuberculosis, adverse
    effects 110–11
Strongyloides stercoralis infection 188,
    234, 255, Plate 19
    autoinfection cycle 197–8
    clinical features 198
    diagnosis 199–201
    epidemiology and control 201
    hyperinfection syndrome 199
    life cycle 197
    treatment 201
Subacute myelo-optic neuropathy
    syndrome 154
Sulfadoxine in malaria 5, 22–3
Sulfalene, malaria 5
Suramin, African trypanosomiasis
    59–60
Swimmer's itch 222
Symmers pipestem fibrosis of
    liver 228
Syphilis, ulcers 327

Taenia saginata 209–10, 211
Taenia solium 210, 211–14
    control 214
    diagnosis 212
    treatment 213–14
Tahlquist papers 284
Tapeworm
    beef 209–10
    hydatid see Hydatid disease
    pork 211–14
Temephos (Abate)
    Cyclops control 274
    Simulium fly control 270
    yellow fever 79
Ternidens deminutus infection 188
Tetrachloroethylene, hookworm
    infection 191–2
Tetracyclines
    African tick typhus 76
    amoebiasis 151

amoebic dysentery 151
balantidiasis 161
cholera 167
relapsing fever 72
scrub typhus 74–5
Tetraethylthiuram monosulphide,
    scabies 328
Thalassaemia 15, 285
Thalidomide, leprosy 304
Thiabendazole
    cutaneous larva migrans 329
    strongloidiasis 199, 201
    toxocariasis 203
Thiacetazone, tuberculosis 107, 108,
    109
    adverse effects 111
    HIV infection 125
Thrombosis, deep venous, typhoid
    fever 97
Tick-borne relapsing fever 70
    endemic 71
Tinidazole, giardiasis 159
Toxocara canis 202
Toxocara cati 202
Toxocariasis
    diagnosis 203
    epidemiology and control 204
    treatment 203
Toxoplasma gondii 137, 204
Toxoplasmosis 137
Travellers' diarrhoea 176
Trichostrongylus infection 188–9
Trichuris infections 142, 175
Trichuris (Trichocephalus) trichiura
    infection 234
    clinical effects 194
    diagnosis 195
    epidemiology and control 196
    life cycle 194
    treatment 195
Trophozoites in malaria 7–8
Tropical eosinophilia syndrome
    275–6
Tropical splenomegaly syndrome 16
Tropical ulcer 327, Plates 30 & 31
    clinical picture and natural
        history 313
    complications 314
    diagnosis 314
    epidemiology and control 316
    treatment 315–16
        inadequate, effects 314

Trypanosoma brucei brucei 50
Trypanosoma brucei gambiense 50, 51
  clinical picture 55–7
  differentiation from *T. b. rhodesiense* 53, 58–9
  epidemiology 62–3
  life cycle 52
  principles of control 62–3
  treatment 59–61
Trypanosoma brucei rhodesiense 50, 51
  clinical picture 57
  differentiation from *T. b. gambiense* 53–4, 58–9
  epidemiology 63–4
  life cycle 52
  principles of control 63–4
  treatment 59–61
Trypanosoma cruzi 67
Trypanosomiasis, African
  clinical picture
    T. b. gambiense infection 55–7
    T. b. rhodesiense infection 57
  CNS involvement 56–7, 59
    treatment 60
  diagnosis 57–9
  disease, course 53–4
  distribution 51
  epidemiology 62–4
  immune response 54
  monitoring cure 62
  parasites 50
    life cycle 50–3
  surveillance and notification 64
  treatment 59–62
    schedule 61
  trypanosomal chancre 53
  For Questions, problems, and cases see pp. 64–6
Trypanosomiasis, South American 67–9
Trypomastigote, African trypanosomiasis 50, 51, 53
Tsetse flies, African trypanosomiasis 50, 51, 62–4
Tsutsugamushi fever 73
Tuberculin test 104
Tuberculoid leprosy 296
Tuberculosis
  antituberculous drugs 108
    adverse effects 110–11
    challenge doses 111
    failure and toxicity rates 109

bone and joint 106
chemotherapy 107–12
  hypersensitivity reactions 111
  short-course 109
clinical syndromes 103–4
drug regimens 109
epidemiology 102–3
fever 136
gastrointestinal and abdominal 106
genitourinary 106
historical aspects 102
HIV 107, 124–5
  chemoprophylaxis 129
  miliary haematogenous dissemination 105
pathogenesis 103–4
pulmonary 104–5
transmission 102–3
treatment failure or relapse 110
Tuberculous lymphadenopathy 105
Tuberculous meningitis 105–6
Tuberculous peritonitis 137
Tuberculous ulcer 327
Tumbu fly 330
Tunga penetrans 330
Typhoid fever 94–101
  clinical picture 95–7
  complications 96–7
  diagnosis 97–8
  disease, course 94–5
  treatment 99–101
    carriers 100–1
Typhus
  African tick 75–6
  American tick 76
  scrub (mite) 73–5

Ulcerative colitis 175
Ulcers 327
  amoebic 142, 144
  Buruli 317–18
  chiclero's 321
  leishmaniasis, cutaneous 323–4
  leprosy 306
  tropical 313–16
  veld sore 327
Urine
  filtration, schistosomiasis 232
  sedimentation, schistosomiasis 231

Vaccines
  antirabies 90
  BCG in leprosy 307
  cholera 168
  cutaneous leishmaniasis 326
  typhoid 101
Veld sore 327
Venezuelan equine encephalitis 83
Verbal autopsies 31
Vibrio cholerae 162
Vincent's organisms, tropical
      ulcer 313
Vipers, bites 335
  treatment 337–8
Virus infections 77–93
  see also Arbovirus infections;
      Haemorrhagic viruses; Rabies
Visceral larva migrans (VLM)   202–4
Visceral leishmaniasis see
      Leishmaniasis, visceral

Washerwoman's hands, cholera 163
Weil–Felix test
  African tick typhus 75
  scrub typhus 75

West Nile virus infection 81
Western blots, HIV testing 115
Western equine encephalitis 83
Whip worm 194
WHO intravenous diarrhoea
      treatment solution, chol-
      era 165
Whole-blood clotting test 337
Widal test 98
Winterbottom's sign 55
Wuchereria bancrofti infection 258,
      275
  clinical picture 251–2
  diagnosis 254–5
  distribution and morphology 251
  epidemiology 256–7
  life cycle 251
  load-related pathology para-
      dox 253–4
  natural history 252–4
  treatment 255–6
  For Questions, problems and cases
      see p. 257

Yellow fever 78–80